Springer Tracts in Civil Engineering

Springer Tracts in Civil Engineering (STCE) publishes the latest developments in Civil Engineering - quickly, informally and in top quality. The series scope includes monographs, professional books, graduate textbooks and edited volumes, as well as outstanding PhD theses. Its goal is to cover all the main branches of civil engineering, both theoretical and applied, including:

- Construction and Structural Mechanics
- Building Materials
- Concrete, Steel and Timber Structures
- Geotechnical Engineering
- Earthquake Engineering
- Coastal Engineering; Ocean and Offshore Engineering
- Hydraulics, Hydrology and Water Resources Engineering
- Environmental Engineering and Sustainability
- Structural Health and Monitoring
- Surveying and Geographical Information Systems
- Heating, Ventilation and Air Conditioning (HVAC)
- Transportation and Traffic
- Risk Analysis
- Safety and Security

Indexed by Scopus

To submit a proposal or request further information, please contact: Pierpaolo Riva at Pierpaolo.Riva@springer.com (Europe and Americas) Mengchu Huang at mengchu.huang@springer.com (China)

More information about this series at http://www.springer.com/series/15088

Hui Liu
Editor

Flood Prevention and Drought Relief in Mekong River Basin

中国水利水电出版社
www.waterpub.com.cn

Springer

Editor
Hui Liu
China Institute of Water Resources
and Hydropower Research
Beijing, China

ISSN 2366-259X ISSN 2366-2603 (electronic)
Springer Tracts in Civil Engineering
ISBN 978-981-15-2005-1 ISBN 978-981-15-2006-8 (eBook)
https://doi.org/10.1007/978-981-15-2006-8

Jointly published with China Water & Power Press
The print edition is not for sale in China. Customers from China please order the print book from: China Water & Power Press.

This Springer imprint is published by the registered company Springer Nature Singapore Pte Ltd.
The registered company address is: 152 Beach Road, #21-01/04 Gateway East, Singapore 189721, Singapore

Preface

This book was conceived as a product of the Lancang-Mekong Cooperation mechanism. The LMC Fund supports work in flood and drought mitigation in the Lancang-Mekong Basin, which is the common concern of the six member countries. This book draws together relative resources and materials on the flood and drought topic of this region, as well as analysis and modelling, so that all member countries and a wider group of people could benefit from them.

The Lancang-Mekong River nurtures the long history and the splendid culture along with it, and is a common precious resource of countries in the basin. Attaching great importance to the integrated management, rational development, and effective protection of water resources, Lancang-Mekong countries are committed to propping up the sustainable development of society and economy through the sustainable utilization of water resources. In March 2016, the First Lancang-Mekong Cooperation (hereinafter referred to as "LMC") Leaders' Meeting was held, marking the formal establishment of the Lancang-Mekong cooperation mechanism, on which the *Sanya Declaration* that stresses water resources cooperation should be a prior field and flagship field was published. In the list of early-harvest projects of LMC, "Strengthening coordinated flood and drought management in the Lancang-Mekong basin" the Vietnamese side proposed was among the two early-harvest projects in the water resources field, and it showcases the six Lancang-Mekong countries are commonly concerned with and attach importance to flood and drought prevention and mitigation. People of all countries in the region will get more benefits if relevant countries step up communication and coordination and carry out practical cooperation under the LMC mechanism.

Affected by tropical monsoons, typhoons, climate change, and El Nino, Mekong River Basin countries are often subject to flood or drought disasters to a varying degree. Effectively preventing and coping with flood and drought disasters is a problem with socio-economic development and human well-being and also a common challenge facing all countries. The Chinese side recognizes that carrying out a joint assessment of flood and drought management together with all basin countries is an important measure for commonly understanding the flood and drought situation in the Mekong River Basin, enhance mutual trust and remove misgivings, cope with flood and drought risks jointly, and build neighborly international relations. In view of this, the Chinese side takes the lead to apply for the "Joint Assessment on the Current Status of Flood and Drought Management in the Mekong River Basin (Phase I)" project as sponsored by the LMC fund. This project is of great significance for bringing benefit to people of Lancang-Mekong countries and promoting regional sustained stability and shared prosperity and development because we will carry out communication and cooperation in water resources field, especially in disaster prevention and mitigation, implement the spirit of the 2nd LMC Leaders' Meeting and the consensus of the Joint Working Group of Lancang-Mekong Water Resources Cooperation, promote the implementation of LMC early-harvest projects proactively, deepen the understanding of the characteristics and current status of flood and drought disasters in the Mekong River Basin, share disaster prevention and mitigation technology, and experience and join hands to cope with flood and drought risks and challenges.

During the First Meeting of the Lancang-Mekong Water Resources Cooperation Joint Working Group (hereinafter referred to as "LMC Water") in February 2017, the Chinese side released to representatives of all countries present at the meeting information about project design, and the project proposal won support from all representatives of countries. In January 2018, the 2nd LMC Leaders' Meeting published the *Five-Year Plan of Action on Lancang-Mekong Cooperation (2018–2022)*, stating the joint assessment of flood and drought management in the Mekong River Basin would be implemented. On April 5, 2018, Mr. Yu Xingjun, Leader of the LMC Water China, sent to the other five countries' leaders of LMC Water a letter giving an introduction to the joint research content and inviting experts to found a joint assessment expert group; in early August, the Chinese expert group carried out, at the invitation of Thailand's leading unit (Ministry of Natural Resources and Environment) and Viet Nam's leading unit (Viet Nam National Mekong Committee), a seven-day field investigation in Thailand and Viet Nam, with in-depth understanding of structural measures and non-structural measures for flood and drought management in Mekong River Basin countries; in mid-August, experts of five Mekong River Basin countries came to China to have a week's technical exchange on current status and technical need of floods and droughts in the Mekong River Basin; in mid-October, Chinese experts formulated the first draft of the joint assessment report and submitted it to the experts of the member countries for discussion and further improvement, and invited them to have technical exchange in China in 22–24 October.

With guidance of the LMC Water, and joint efforts of experts from all LMC member countries, the joint assessment expert group finally formed a report with general consent of the LMC member countries, and ready for sharing and publication in early 2019.

Beijing, China

Hui Liu

Acknowledgements

This book is a team effort at its best. The work was supported by the Joint Working Group on Lancang-Mekong Water Resources Cooperation under the Lancang-Mekong Cooperation mechanism. We are more grateful than we can say for contributions of colleagues from Lancang-Mekong Cooperation member countries and its line agencies, who provided relentless help and valuable suggestions. We especially owe thanks to Hlaing Tun, Thongthip Chandalasan, Winai Wangpimool, Chea Monyvycet, Nguyen Dinh Dat, Toe Toe Aung, Aung Than Oo, Somphone Khamphanh, Phaylin Bouakeo, Prasith Deemanivong, Wasna Roi-Amphaeng, and Supapap Patsinghasanee.

The production of this book would not have been possible without the support of China Water & Power Press. We would like to thank Ms. Lijuan Xu for copyediting and coordinating the publication processes.

Disclaimer

This book was prepared by the China Institute of Water Resources and Hydropower Research (IWHR) and Tsinghua University. Although all effort is made to ensure that the information, views, and analyses contained in this document are based on sources believed to be reliable, no representation, expressed or implied, is made guaranteeing accuracy, completeness or correctness. The opinions contained herein reflect the judgment of the authors and are subject to change without notice. Neither the IWHR nor the Tsinghua University, the Lancang–Mekong Cooperation Member Countries and any agency thereof, their employees, contractors, subcontractors or their employees, assumes any legal liability or responsibility for the consequences of any third party's use of the information, opinions, and analysis contained in this document.

Contents

Chapter 1
Overview of the Mekong River Basin

Baiyinbaoligao, Hui Liu, Xingru Chen, and Xiangpeng Mu

Abstract The Lancang-Mekong River, the largest transboundary river in Southeast Asia, flows through six riparian countries, nurtures splendid culture along it. The river is called Lancang River in China, and Mekong River out of China. The Mekong River region is featured by various landscape patterns and high variability of rainfall caused by monsoon cycles. This region is also featured by high density of population, high potential of development and high international attention, which make this study more meaningful. The physical geographical feature of this region is illustrated from the aspects of topography, meteorology and hydrology in this chapter. The Socio-economy characteristics of this region is also reviewed in this chapter to provide background knowledge for the further analysis of flood and drought management. The irrigation projects and hydropower projects, main types of structural measures, are also briefly introduced from the whole basin view.

1.1 The Lancang-Mekong River Basin

The water resources challenges of the Southeastern Asia regions is exceptionally complex, with growing populations and economies needing reliable and predictable water resources for livelihoods and food and energy production. The region is characterized by monsoon cycles leading to exceptionally high rainfall and runoff variability, and extremes of severe flooding and drought, overlain with uncertainties of

Baiyinbaoligao (✉) · H. Liu · X. Chen · X. Mu
China Institute of Water Resources and Hydropower Research, A-1 Fuxing Road, Haidian District, Beijing 100038, China
e-mail: baiyin@iwhr.com

H. Liu
e-mail: wuyouliuhui@163.com

X. Chen
e-mail: chenxr@iwhr.com

X. Mu
e-mail: swood2002@163.com

H. Liu (ed.), *Flood Prevention and Drought Relief in Mekong River Basin*, Springer Tracts in Civil Engineering,
https://doi.org/10.1007/978-981-15-2006-8_1

a changing climate. Many great rivers rise from the Himalayas and are shared by more than one nation, with almost all of the region's mainland nations dependent to a greater or lesser extent on these transboundary waters. Most of the flow of these rivers comes from midstream precipitation, with glacier melt providing less than 20% of these rivers' overall flow, except in the Indus basin (c. 50%).[1]

Lancang River originates from Yushu Tibetan Autonomous Prefecture, Qinghai Province, China, and is known as Mekong River[2] after flowing out of China from Xishuangbanna Dai Autonomous Prefecture, Yunnan Province. Mekong River flows through 5 countries including Myanmar, Lao PDR, Thailand, Cambodia and Viet Nam into South China Sea from the west of Ho Chi Minh City, Viet Nam, as shown in Fig. 1.1.

The Lancang-Mekong River runs as long as 4880 km, with a drainage area of 795,000 km^2[3] and an average annual runoff volume of 475 km^3. It is the tenth longest river and ranks eleventh in terms of annual water yield in the world, and is amongst the seven biggest rivers in Southeast Asia. In specific, Mekong River registers a drainage area of 630,600 km^2, a length of 2750 km and an average annual runoff volume of 410.9 km^3, taking up 79%, 56% and 86.5% respectively of Lancang-Mekong River. From the source in Qinghai-Tibet Plateau to the estuary, the river flows through a slope of 5060 m, with an average gradient of 1.04‰.[4]

Distribution of area and river length of the Lancang-Mekong River Basin by country is demonstrated in Table 1.1.

1.1.1 Lancang River Basin

About half (57%) of the total length of the Lancang-Mekong River of 4884 km is located in the territory of China. Lancang River basin is mainly steep alpine valley, located in the under-developed region with extremely inconvenient transportation and deficient natural resources, except extraordinary rich hydropower resources. Water utilization rate is about 3% in this region, and the water consumed is less than 1% of the total runoff of Lancang-Mekong basin. This zone contributes about

[1]Tian Fuqiang and Liu Hui. China-shared rivers, shared futures. Vientiane Times, pp 15–16. July 14, 2016.

[2]In most MRC publications, it is usually called as Lower Mekong River. To keep insistent with the former joint assessment carried out by MWR, China and MRCS in 2016, we use Mekong River instead of Lower Mekong River. As to the Upper, Middle and Lower reaches of Mekong River, Nong Khai and Kratie hydrological station is choosen as the division point for the three sections of the mainstream Mekong River.

[3]The total Lancang-Mekong Basin area of 795,000 km^2 is used in MRC publications (e.g. Overview of the Hydrology of the Mekong Basin), however, China suggests the total Lancang-Mekong Basin area of 812,400 km^2.

[4]Mekong River Commission and Ministry of Water Resources of the People's Republic of China (2016). Technical Report—Joint Observation and Evaluation of the Emergency Water Supplement from China to the Mekong River. Mekong River Commission, Vientiane, Lao PDR.

Fig. 1.1 The sub-basins of the Mekong River Basin

Table 1.1 Distribution of the area and river length of Lancang-Mekong River Basin by Country [Tang Haixing, Water resources in the Lancang-Mekong River Basin and analysis on the present situation of its utilization. Yunnan Geographic Environment Research, 1999, 11(1), 16–2 (in Chinese)]

	China	Myanmar	Lao PDR	Thailand	Cambodia	Viet Nam	Total
Drainage area/10^4 km^2	16.5	2.4	20.2	18.4	15.5	6.5	79.5
% of total drainage area	21	3	25	23	20	8	100
% of national territorial area	1.7	3.6	85.2	35.9	85.6	19.7	–
Length (inland)/km	2130		777		502	230	3639
Length (boundary)/km	31		234	976	–	–	1241

13.5% runoff of Lancang-Mekong River. The runoff comes from rainfall, snowmelt and groundwater seeping. This region has a distinguishing rainy season and dry season, the runoff of rainy season accounts for 75% of the annual runoff of Lancang River. The dry season lasts from November to April, during which the runoff mainly depends on snowmelt and groundwater seeping.

The construction of cascade reservoirs in the mainstream in the middle and lower reaches of the Lancang River has been completed. Xiaowan Reservoir and Nuozhadu Reservoir have especially the multi-year regulating capacity, with regulating storage of 21.2 billion m^3 in total. By operating and regulating scientifically, Lancang River cascade reservoirs are capable to balance the water discharge/volume between the wet season and dry season, benefiting the Mekong River on the aspects of flood control, irrigation, navigation and so on.[5]

1.1.2 Mekong River Basin

Locations and area of the primary sub-tributaries in the Mekong River Basin are shown in Fig. 1.1 and Table 1.2.[6] Main tributaries include Nam Tha, Nam Ou, Nam Kam, Nam Ngum, Nam Cadinh, Se Bang Fai, Se Bang Hieng, Nam Mun, Nam Sang and Tonle Sap River, and the Nam Mun is the biggest tributary.

[5]Mekong River Commission and Ministry of Water Resources of the People's Republic of China (2016). Technical Report—Joint Observation and Evaluation of the Emergency Water Supplement from China to the Mekong River. Mekong River Commission, Vientiane, Lao PDR, p 2.

[6]The first-class tributaries are drawn with reference to the MRC publications. Table 1.2 is the corresponding statistical area of the first class tributaries. The name of the tributaries is determined by that of the rivers converge on the Mekong mainstream.

Table 1.2 Sub-basin area of Mekong River Basin [Mekong R C. Planning Atlas of the Lower Mekong River Basin. Phnom Penh: Lao Uniprint Press Co. Ltd., 2011, p 9 (Though most of the primary tributaries were referred to this publication, the author did not split larger tributaries like Tonle Sap into sub-basins as the publication did)]

Name	Area (km^2)	Name	Area (km^2)	Name	Area (km^2)
Ban Khai San	778	Nam Khan	7490	Nam Thong	455
Ban Nam Song	138	Nam Khop	1521	Nam Ton	587
Delta	48235	Nam Loei	4012	O Talas	1448
Doi Luang Pae Muang	688	Nam Ma	1141	Phu Luong Yot Huai Dua	491
Huai Khok	538	Nam Mae Ing	7267	Phu Pa Huak	132
Huai Bang Bot	2402	Nam Mae Kham	4079	Prek Chhlong	5957
Huai Bang Koi	3313	Nam Mae Kok	10701	Prek Kamp	1142
Huai Ma Hiao	990	Nam Mae Ngao	485	Pre Krieng	3332
Huai Nam Huai	1755	Nam Mang	1836	Prek Mun	476
Huai Sophay	186	Nam Mang Ngai	944	Prek Preah	2400
Huaag Hua	626	Nam Mi	1032	Prek Te	4364
Huai Bang Haak	938	Nam Mun	70574	Prek Thnot	6124
Huai Bang I	1496	Nam Nago	1008	Se Bang Fai	10407
Huai Bang Lieng	695	Nam Ngam	489	Se Bang Hieng	19958
Huai Bang Sai	1367	Nam Ngaou	1495	Se Bang Nouan	3048
Huai Ho	691	Nam Ngeun	1819	Se Done	7229
Huai Khamouan	3762	Nam Ngum	16906	Se Kong	28815
Huai Luang	4090	Nam Nhah	316	Se San	18888
Huai Mong	2700	Nam Kadun	456	Nam Sing	2681
Huai Muk	792	Nam Kai	602	Nam Songkhram	13123
Huai Nam Som	1072	Nam Kam	3495	Nam Suai	1247
Huai Som Pak	2516	Nam Keung	633	NamSuong	6578
Huai Thuai	739	Nam Nhiep	4577	Nam Tam	1548
Huai Tomo	2611	Nam Nuao	2287	Nam Tha	8918
Muang Liep	488	Nam Ou	26033	Nam Thon	838
Nam Beng	2131	Nam Pho	2855	Siem Bok	8851

(continued)

Table 1.2 (continued)

Name	Area (km²)	Name	Area (km²)	Name	Area (km²)
Nam Cadinh	14822	Nam Phone	664	Sre Pok	30942
Nam Chi	49133	Nam Phoul	2095	Tonle Repon	2379
Nam Heung	4901	Nam Phuong	4139	Tonle Sap Lake	85850
Nam Hinboun	2529	Nam Sane	2226		
Nam Houng	2872	Nam Sang	1290		

(1) **Chinese border—Chiang Saen, Thailand**

From the Chinese border to Chiang Saen, the Mekong river section is mainly the boundary river between Lao PDR and Myanmar, and between Lao PDR and Thailand. Mainly made up of canyon rivers, the river reach gets its runoff largely from the Lancang River in China and several small tributaries in Thailand, Lao PDR and Myanmar. Tributaries on the right bank of the Mekong River within the territory of Myanmar are the Nam Loi, Nam Yang, Nam Leung and Nam Mae Kham, involving a drainage area of 28,600 km^2 and an average annual runoff volume of 17.63 km^3. Tributaries on the left bank of the Mekong River within the territory of Lao PDR include the Nam Ma, Nam Pho and Nam Nuao. Part of the Nam Mae Kham is located on the right bank of the Mekong River within the territory of Thailand.

(2) **Chiang Saen-Vientiane**

From Chiang Saen to Vientiane, the mainstream Mekong River begins to be wider with ladder-like floodplains and paddy and other crop fields. On Thailand's side in Chiang Saen are two big tributaries, the Nam Mae Kok and the Mae Nam Ing, and such big tributaries as the Nam Tha, Nam Ou, Nam Soung and Nam Khan, which originate from the Annam Mountainous Region, are located near Luang Prabang in Lao PDR.

(3) **Vientiane-south frontier of Lao PDR**

On the right bank in Thailand are the Huai Lang River, Huai Luang River, Nam Songkhram, Nam Mum and other tributaries; and tributaries originating from the Annam Mountainous Region which are located on the left bank in Lao PDR include the Nam Ngum, Nam Cadinh, Nam Hinboun, Se Bang Fai, Se Bang Hieng and Se Done.

(4) **Within the territory of Cambodia**

Tributaries on the right bank within the territory of Cambodia mainly include the Tonle Sap River and the Tonle Repon River. In specific, the Tonle Sap River has a catchment area of about 85,850 km^2 (including lake area), and at its upstream is the Tonle Sap Lake that not only stores and retains floodwater from upper mountainous regions but also stores and retains a large amount of floodwater of the Mekong River in the wet season and replenishes it back to the Mekong River during the dry season.

On the left bank are three big tributaries, Se Kong, Se San and Sre Pok, which originate from the Annam Mountainous Region in Lao PDR and Viet Nam. Their catchment area amounts to 78,645 km^2, taking up roughly 13% of the drainage area of the Mekong River, and their water volume makes up 23% of the average runoff of the basin.

(5) **Delta region**

Also known as the Cuu Long River Delta, the Mekong Delta is situated at the southernmost point of Viet Nam and the southeastern part of Cambodia. With an area of 48,235 km^2, it is the largest plain in Southeast Asia. After flowing through Phnom Penh, the Mekong River gets divided into two rivers known as Tien Giang and Hau Giang within the territory of Viet Nam and then, divided into six rivers. Because of separation by sandbars, there are nine estuaries leading to the sea.

1.2 Physical Geography

1.2.1 Topography

The upper Mekong River covers the east of Shan State of Myanmar, the north of Lao PDR and the north of Thailand. Mostly covered with hills and mountains, the east of Shan State, Myanmar is a hilly and mountainous area. It is the central zone of Golden Triangle. The north of Lao PDR is covered with plateaus and low mountains including Xiangkhoang Plateau, Tran Ninh Plateau, Sipsongchutai Range, Luang Prabang Range and Phetchabun Range. Mountains, hills and plateaus take up 85% of its area and inter-mountainous plains, flat lands and basins only take up 15%. Its altitude is between 1000 and 2000 m. The north of Thailand borders Luang Prabang Mountainous Area to the east. Dean Lao Mountainous Region is located in the northwest, and middle part is a broad plain. The average altitude is about 400 m. This area is also a part of Golden Triangle.

The middle reaches of Mekong River include the central and southern parts of Lao PDR and the northeastern parts of Thailand. The topography is complicated, including Phou Luang Range, Khammouane Plateau, Bolovens Plateau, Vientiane Plain, Thakhek Valley, Savannakhet Plain and Champasak Lowland in central and southern Lao PDR. In northeastern Thailand are Khorat Plateau, Phu Phan Mountains, Buri Ram Hills and Ubon Ratchathani Steppe. In addition to plateaus and low mountains, the left bank of Mekong River is composed of plains and lowlands.

The lower Mekong River covers Cambodia and most parts of southern Viet Nam, with high terrains in three directions and low terrains in the middle part. Eastern, northern and western parts are covered with plateaus and mountains. Most Cambodian areas in this region are plains, and central plains take up 46% of the total area of Cambodia. Located in the northeast of the central plains, the Truong Son Range separates Lancang-Mekong River from rivers flowing to the South China Sea. Situated in the west and the southwest, the Cardamom Mountainous Region is a watershed between Mekong River and rivers flowing to the Gulf of Siam. The Dangrek Mountains constitute a natural border between Cambodia and Thailand. Cambodia has 1,200,000 hm^2 and Viet Nam takes up 3,900,000 hm^2 in Mekong Delta. In Viet Nam, Mekong Delta consists of Tien Giang Plain, Hau Giang Plain and Dong Thap Muoi Plain.

Topography of the Mekong River Basin is indicated in Fig. 1.2.

1.2.2 Meteorology

Situated in the tropical monsoon region of Asia, the Mekong River Basin has distinct wet and dry seasons under the influence of the monsoon climate, with extremely uneven annual distribution of precipitation. The Southwest Monsoon starts from May until the late September, and then the Northeast Monsoon lasts from November to mid-March. As the Southwest Monsoon brings a lot of moisture from the sea to coastal countries, abundant rainfall is enjoyed, especially in hilly and mountainous areas. On the contrary, the Northeast Monsoon comes from mainland, so it is dry and rainfall is rarely witnessed. Precipitation is uneven between seasons, which is very evident between wet and dry seasons. 80% of precipitation gathers in the wet season between May and October.

In the meantime, the spatial distribution of rainfall in the Mekong River Basin is also very uneven, as shown in Fig. 1.3. Under the influence of terrain, movement of the Southwest Monsoon, the left bank of Mekong River is multi rain belt, with much higher runoff yield than the right bank, it takes up 70% precipitation of the whole basin. Annual precipitation rises from 1000 mm in northeastern Thailand to 4000 mm in the mountainous borders of the basin within Lao PDR, Cambodia and Viet Nam. In the Chi-Mun River Basin of Thailand and the Tonle Sap Lake Basin of eastern Cambodia, average annual precipitation declines evidently from east to west.

Temperature change is even within the Mekong River Basin, with an average annual temperature between 25 and 27 °C. Average relative atmospheric humidity is the highest, slightly higher than 80%, in September, and the lowest, merely 60%, in March.

1.2.3 Hydrology

Hydrology closely reflects the pattern in rainfall distribution described in Fig. 1.3, but also affected by topography, vegetation and soil texture. Accordig to MRC publication, the flow contribution from tributaries to the mainstream Lancang-Mekong River in various reaches is shown in Table 1.3. The flood season in the Mekong River Basin lasts from June to November and accounts for 80–90% of the total annual flow.

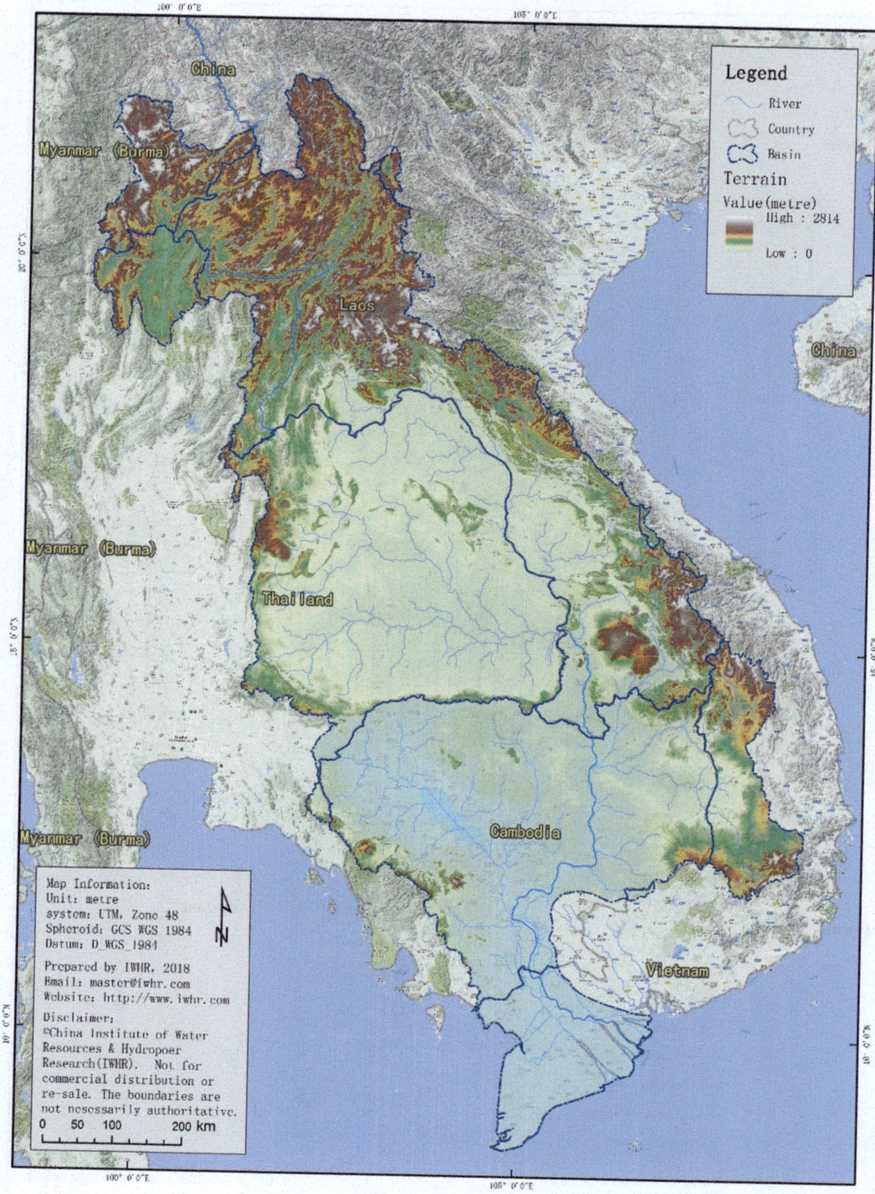

Fig. 1.2 Topography of the Mekong River Basin

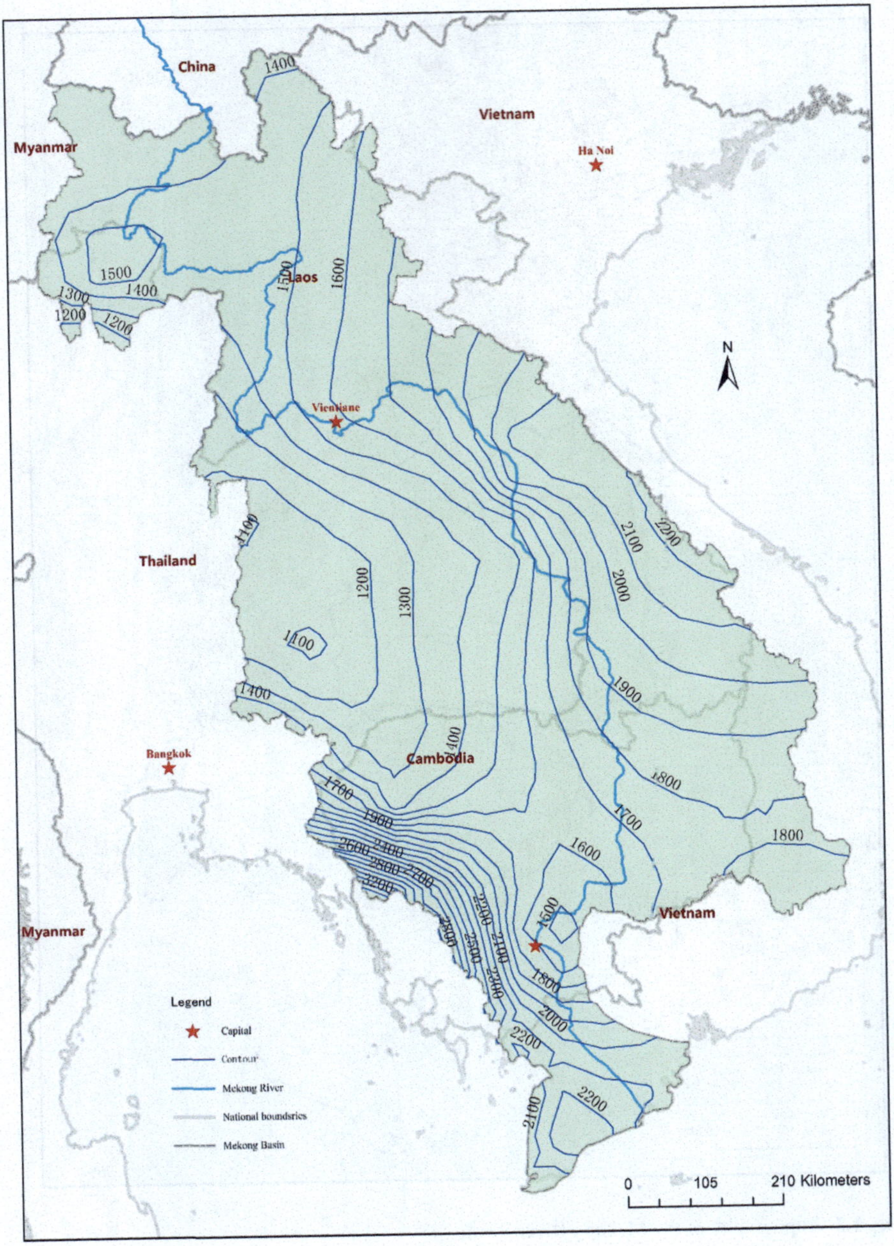

Fig. 1.3 Annual precipitation distribution of the Mekong River Basin [Based on monthly precipitation data (1901–2016) from Climate Research Unit (CRU). The unit is mm]

Table 1.3 Proportional contributions to total Lancang-Mekong River mean annual flow by river reach, distinguishing those made by the left and right bank tributary systems (Mekong R C. Planning Atlas of the Lower Mekong River Basin. Phnom Penh: Lao Uniprint Press Co. Ltd., 2011, p 53)

River reach	Left bank (%)		Right bank (%)	Total (%)
Lancang	16[a]			16[a]
Lancang–Chiang Saen	1		4	5
Chiang Saen–Luang Prabang	6		3	9
Luang Prabang–Chiang Khan	1		2	3
Chiang Khan–Vientiane	0		0	0
Vientiane–Nong Khai	0		1	1
Nong Khai–Nakhon Phanom	19		4	24
Nakhon Phanom–Mukdahan	3		1	4
Mukdahan–Pakse	5		6	11
Pakse–Stung Treng	23		3	26
Stung Treng–Kratie	1		0	1
Total	60	16	24	100

[a]China suggests the proportional contribution of Lancang River is 13.5% [Mekong River Commission and Ministry of Water Resources of the People's Republic of China (2016). Technical Report—Joint Observation and Evaluation of the Emergency Water Supplement from China to the Mekong River. Mekong River Commission, Vientiane, Lao PDR, p 2]

The wet and steep left bank tributaries used to have high runoff depth.[7] Nam Ngun and Nam Kading-Nam Thuen produce the greastest runoff depth up to 2500 mm per year, accounting for around 5% of the Lancang-Mekong total runoff for each catchment. The left bank tributaries between Vientiane and Nakhon Phanom contribute around 19% of the total runoff. The 3 s River (Se Kong-Se San-Sre Pok), entering the mainstream between Pakse and Stung Treng, contritutes 17% of the total runoff.

The drier and flatter catchements on the right bank used to have low runoff depth. Among them, Chi-Mun River Basin and Tonle Sap Basin have the largest catchment areas, contributing around 5% to the total runoff.

1.3 Socio-economy

1.3.1 Administrative Division

Provincial-level administrative divisions of countries within the Mekong River Basin are shown in Fig. 1.4 and Table 1.4.

All of the 17 provinces of Lao PDR fall within the Mekong River Basin, take up 85.2% of the total area of Lao PDR.

[7]Runoff depth is calculated by dividing the runoff (discharge) by its catchment area.

Fig. 1.4 Administrative divisions of the Mekong River Basin

Table 1.4 Provinces of the 5 Countries within the Mekong River Basin [Mekong R C. Planning Atlas of the Lower Mekong River Basin. Phnom Penh: Lao Uniprint Press Co. Ltd., 2011, p 5. (Most of the information was referred to this publication except that of Myanmar)]

Cambodia	Lao PDR	Myanmar	Thailand	Viet Nam
Banteay Meanchey	Attapeu	Most parts of Kengtung Almost all parts of Mongphat Tarchilaik A small part of Mongsat	Amnat Charoen	An Giang
Battambang	Bokeo		Burirum	Bac Lieu
Kampong Cham	Borikhamxay		Chaiyaphum	Ben Tre
Kampong Chhnang	Champasak		Chantaburi	Binh Phuoc
Kampong Speu	Huaphanh		Chiang Mai	Ca Mau
Kampong Thom	Khammuane		Chiang Rai	Can Tho
Kampot	Luangnaumtha		Katasin	Dak Lak
Kandal	Luangprabang		Khon Kaen	Dak Nong
Koh Kong	Oudomxay		Loei	Dien Bien
Kratie	Phongsaly		Maha Sarakham	Dong Thap
Mondul Kiri	Saravane		Mukdahan	Gia Lai
Otdar Meanchey	Savannakhet		Nakhon Ratchasima	Hau Giang
Pailin	Sekong		Nong Bua Lamphu	Kon Tum
Phnom Penh	Vientiane		Nong Khai	Lam Dong
Preah Sihanouk	Vientiane Capital		Phayao	Long An
Preah Vihear	Xayaboury		Phet Chabun	Quang Tri
Prey Veng	Xiengkhuang		Roi Et	Soc Trang
Pursat			Sa Keo	Thua Thien Hue
Ratanak Kiri			Sakon Nakhon	Tien Giang
Siem Reap			Si Saket	Tra Vinh
Stung Treng			Surin	Vinh Long
Takeo			Ubon Rachathani	
Tonle Sap Lake			Udon Thani	
			Yasothon	
Total				
23	17	4	25	22

All provinces, except Koh Kong, of Cambodia, are wholly or partly contained in the Mekong Basin, accounting for 85.6% of the total national area.

25 out of the 77 provinces of Thailand located in the Mekong Basin, accounting for 35.9% of the total national area.

22 of the 63 provinces of Viet Nam fall in the basin, of which 13 gathered in the Mekong Delta and another 5 located at the sources of Se San Basin and Se Prok Basin, accounting for 19.7% of the total national area.

The area of Myanmar falling whthin the basin only accounts for 3% of the national territory. 4 cities of Shan State locate in the basin, of which Taichilaik wholly and the other 3 partly falling within the basin.

1.3.2 *Population Distribution*

Population of the Mekong River Basin gather largely in Cambodia, Lao PDR, Thailand and Viet Nam, and only a small population live in Myanmar, as shown in Fig. 1.5. Table 1.5 demonstrates the share of each country's population in the basin and the share of the basin population in each country. Population distribution is uneven. Cambodia and Lao PDR only make up 28% of the basin's population though most of their territory is located there. Thailand has 37% of its territorial area located within the Mekong River Basin, and also takes up 37% of its population. 20% of Viet Nam's territorial area is located in Mekong Delta and highland in the middle part, and the country takes up 34% of the basin's population. 80% of the basin's population live in rural areas.

Population density varies greatly across the Mekong River Basin. Mekong Delta within Viet Nam and the area around Phnom Penh, Cambodia have a dense population. In Mekong Delta, average population density is 443 people/km^2. That of Phnom Penh reaches 41,200 people/km^2. Area around Vientiane has a population density of 739 people/km^2. Reasons behind the dense population in these regions include the convenient transport of goods on land and in inland waters, the adjacency with Hanoi, fertile land, and being close to irrigation facilities and waters. In contrast, population density along the highlands bordering Viet Nam is 20 people/km^2, and population density is inversely proportional to altitude.[8]

1.3.3 *Land Utilization*

Land utilization of the basin in 2015 is demonstrated in Fig. 1.6.[9] The main land utilization include forest and paddy field. Paddy rice makes up 22% of the total

[8]Mekong R C. Planning Atlas of the Lower Mekong River Basin. Phnom Penh: Lao Uniprint Press Co. Ltd., 2011, p 15.

[9]Based on landuse database of Tsinghua University. http://data.ess.tsinghua.edu.cn/.

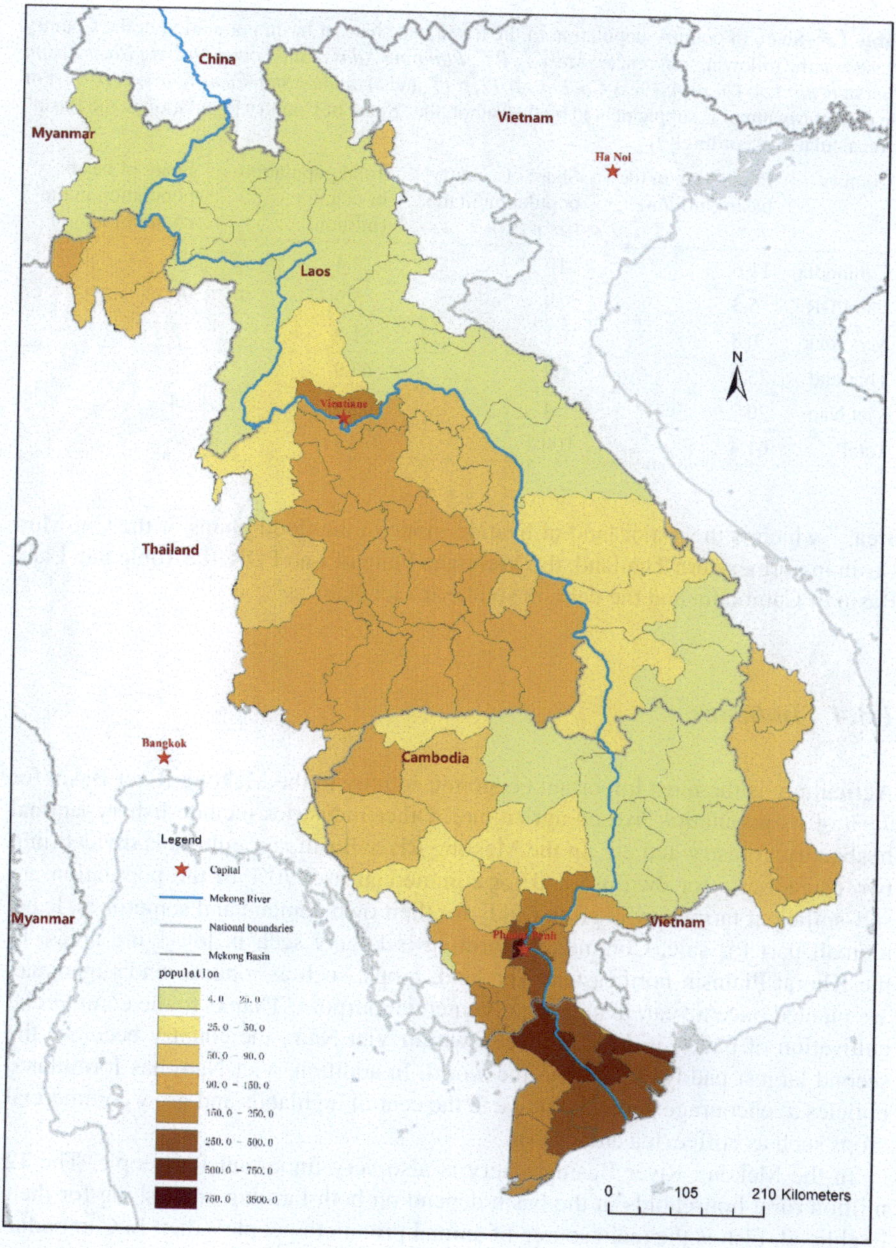

Fig. 1.5 Population distribution of the Mekong River Basin, 2011 [Based on GWPv4 data, Center for International Earth Science Information Network (CIESIN), Columbia University. 2018. Documentation for the Gridded Population of the World, Version 4 (GPWv4), Revision 11 Data Sets. Palisades NY: NASA Socioeconomic Data and Applications Center (SEDAC). https://doi.org/10.7927/H45Q4T5F Accessed 15 July 2019]

Table 1.5 Share of country population in the Basin and share of basin population in the Country [Based on the following references: *Mekong R C. Planning Atlas of the Lower Mekong River Basin. Phnom Penh: Lao Uniprint Press Co. Ltd., 2011, p 13* and *Myanmar Statistical Yearook 2016.* (For the data of Myanmar is supplemented by the author, the "Share of Country Population in the Basin" is recalculated accordingly)]

Country	Population in the basin (million)	Share of country population in the basin (%)	Total population of country (million)	Share of basin population in the country (%)
Cambodia	11.6	19	14.4	81
Lao PDR	5.3	9	5.9	90
Myanmar	0.8	1	51.5	2
Thailand	23.0	37	63.9	36
Viet Nam	20.7	34	87.4	24
Total	61.4	100	223.1	28

area,[10] which is the major land utilization mode in the flood plains of the Chi-Mun Basin in northeastern Thailand, the Vientiane Plain of Lao PDR, the Tonle Sap Lake Basin of Cambodia and the delta in southern Viet Nam.

1.3.4 Industry

Agriculture is the most important economic activity in the Mekong River Basin for 75% of its population live on agriculture. Other industries include fishery, animal husbandry, forestry and etc. In the Mekong River Basin, agriculture is divided into two categories—for own use and for commercial use. 70% of the population are self-sufficient farmers who grow paddy for their own demand and sometimes, leave a small part for sale. Commercial farming is largely seen in low-lying areas. In the Khorat Plain in northeastern Thailand, crops, such as tobacco and sugarcane, are planted once a year, mainly for commercial purpose. Thanks to the commercial cultivation of paddy in Mekong Delta within Viet Nam, the country becomes the second largest paddy exporter in the world. In addition, Viet Nam has formulated policies to encourage people to move to the central highlands and grow commercial crops such as coffee, tea and rubber.

In the Mekong River Basin, fishery is also very important for people. The 12 million rural households in the basin depend on both farming and fishing for their livelihood. Fish is the main source of animal protein in people's diet. In Cambodia, more than 1.2 million people live in near the Tonle Sap Lake, where fishery is the only source of income. The annual fishery output of Tonle Sap Lake in Cambodia is

[10] http://portal.mrcmekong.org/tech_report.

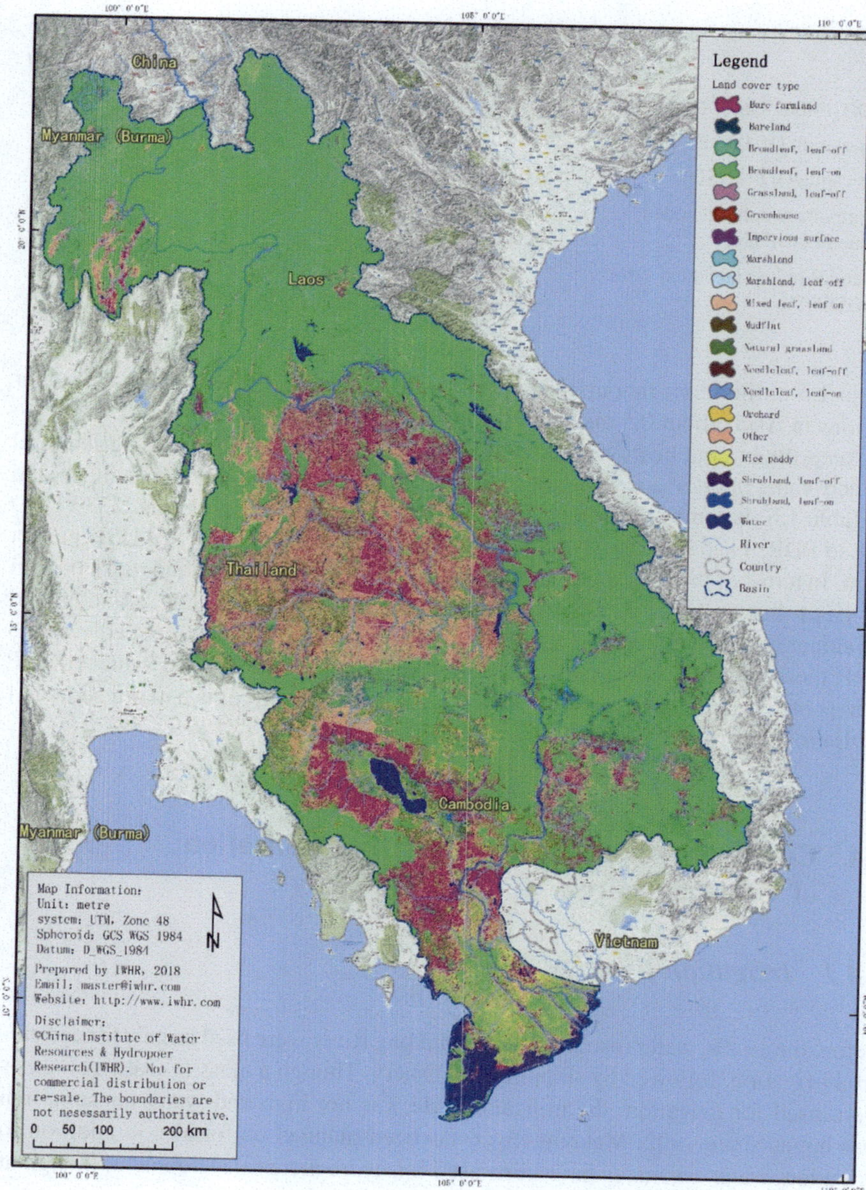

Fig. 1.6 Land utilization of the Mekong River Basin, 2015

Table 1.6 Economic and industrial status of countries in the basin, 2016

	Cambodia	Lao PDR	Myanmar	Thailand	Viet Nam
GDP (billion USD)	200.2	158.1	632.3	4117.6	2052.8
Population (10,000 People)	1576	676	5289	6886	6457
Per capita GDP (USD)	1270	2339	1196	5980	3179
Share of agricultural added value in GDP (%)	25	17	25	9	16
Share of industrial added value in GDP (%)	29	29	35	36	33

about 1.5 million tons (about 230 kg/hm^2), which is much higher than that of other regions in Asia (about 0.5 million tons, that is 100 kg/hm^2).[11]

According to publications of World Bank, per capita GDP, Share of Agricultural Added Value in GDP and Share of Industrial Added Value in GDP are as illustrated in Table 1.6. It can be seen per capita GDP is 1270 USD in Cambodia, 2339 USD in Lao PDR, 1196 USD in Myanmar, 5980 USD in Thailand and 3179 USD in Viet Nam. In terms of the ratio of agricultural added value to GDP, Thailand ranks bottom by 9%, followed by Lao PDR by 17% and Viet Nam by 16%, and Cambodia and Myanmar rank top either by 25%. In terms of the ratio of industrial added value to GDP, these countries have small differences, and the industrial added value-to-GDP ratio is bigger than the agricultural added value-to-GDP ratio. It is a major source of GDP growth for these countries.

1.4 Current Status of Development and Utilization of Water Resources

1.4.1 Irrigation Projects

Before the 1980s, water resources of the Mekong River were used largely for agricultural irrigation, followed by shipping and fishery. Though a mass of water resources were used for irrigation, the utilization rate was not high and water consumption was huge. Since 1992, Mekong River has been planned uniformly; in addition to irrigation, water resources have been used for hydroenergy development, though at a low rate.

[11] MRC Annual Flood Report 2005. Mekong River Commission, Vientiane, Lao PDR. 82 pp. ISSN 1728 3248, p 10.

Cambodia has registered an irrigation area of 479,762 hm^2 in wet season, and will increase it to 774,000 hm^2 as of 2030 and to 838,000 hm^2 as of 2060. Lao PDR has 3094 irrigation projects that cover an area of 225,446 hm^2 in wet season, less than that of other countries in the basin. Thailand has built 134 irrigation projects in the Mekong River Basin. Those projects cover an area of 903,946 hm^2 in wet season, and are mainly located on both banks of rivers and the flood plains in northern Thailand. Viet Nam has registered an irrigation area of 2.38 million hm^2 in the wet season, and won't increase it in next decades. According to MRC publications, existing projects and irrigation area in the Mekong River Basin are shown in Figs. 1.7, 1.8 and 1.9.

Viet Nam has the biggest existing irrigation area, taking up 59.5% of the gross irrigation area of the Mekong River Basin in wet season; following it is Thailand that accounts for 22.5%; the rest 18% is situated in Cambodia and Lao PDR. In the following decades, the irrigation area of Cambodia and Lao PDR in the Mekong River Basin may increase.

1.4.2 Hydropower Projects

The United Nations and the World Bank give high priority to the development of sustainable hydropower, which is deemed low carbon and renewable, Hydropower currently comprises 80% of the World's installed renewable energy, and is a key energy option for both poverty reduction and the reduction of greenhouse gas emissions.[12] Most industrial nations, including China, have developed a relatively high proportion of their hydropower potential. However, China has only developed 10% of its hydropower potential on its transboundary rivers.

Table 1.7 shows hydropower projects in the Mekong River Basin. 11 cascade hydropower station projects are planned on the mainstream Mekong River, of which 2 are located on the boundary between Lao PDR and Thailand, 7 within the territory of Lao PDR and the rest 2 within the territory of Cambodia. In hydropower projects on tributaries of the Mekong River, 91 are located within the territory of Lao PDR, 15 in Viet Nam, 12 in Cambodia and 7 in Thailand.

In Cambodia, 2 mainstream hydropower stations, and 11 tributary dams which are all located in the Se Kong-Se San-Sre Pok Basin, are planned in the Mekong River Basin. In Lao PDR, there are 100 hydropower stations (73.5% of total hydropower stations in the basin) existing, under construction or planned in the Mekong River Basin, of which 9 hydropower stations are located on mainstream and the rest 91 on tributaries. In the Mekong River Basin in Thailand, 7 hydropower stations have been

[12]http://www.un.org/wcm/webdav/site/sustainableenergyforall/shared/Documents/SEFA-Action%20Agenda-Final.pdf; http://www.worldbank.org/en/topic/hydropower/overview#1.

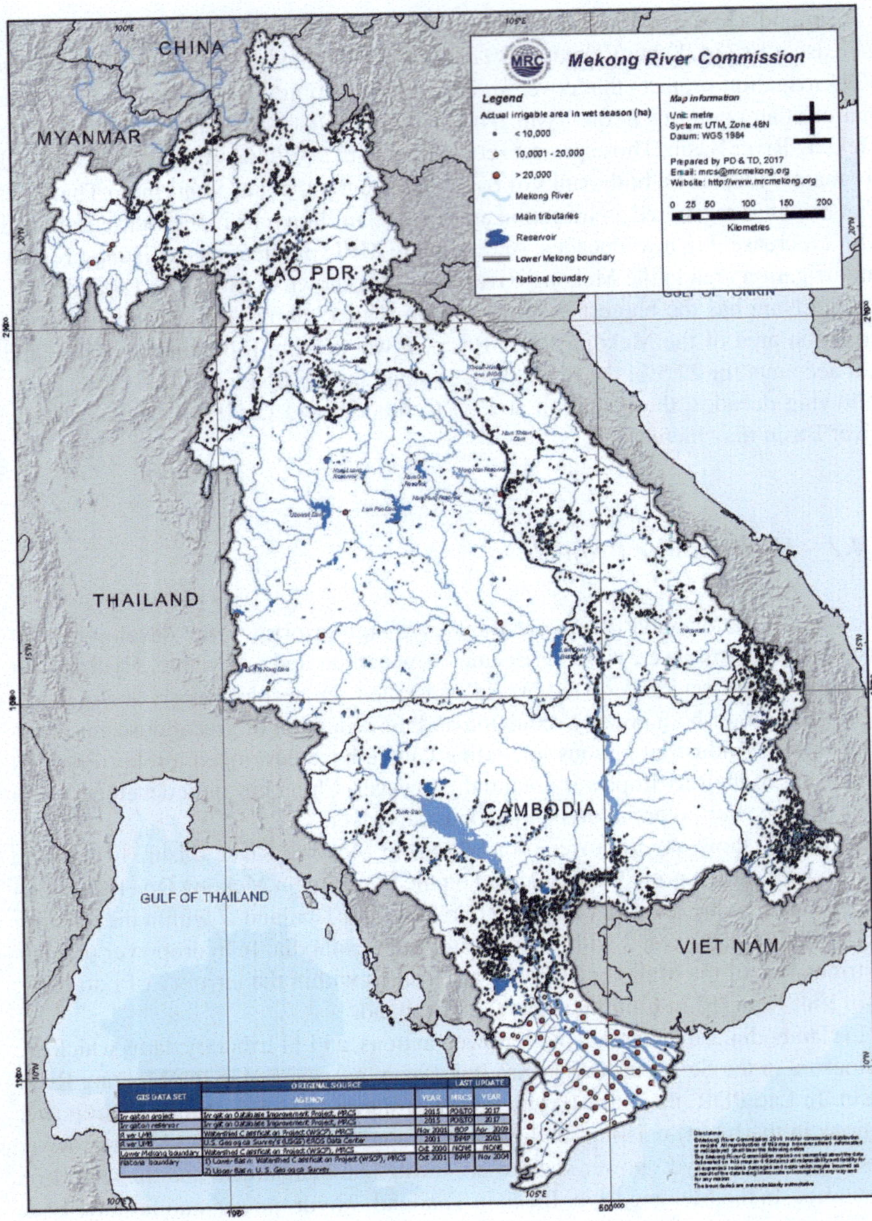

Fig. 1.7 Irrigation projects and existing irrigation reservoirs in the Mekong River Basin (http://www.mrcmekong.org/assets/Publications/MRC-Irrigation-Database-Improvement-final-for-web.pdf, p 53)

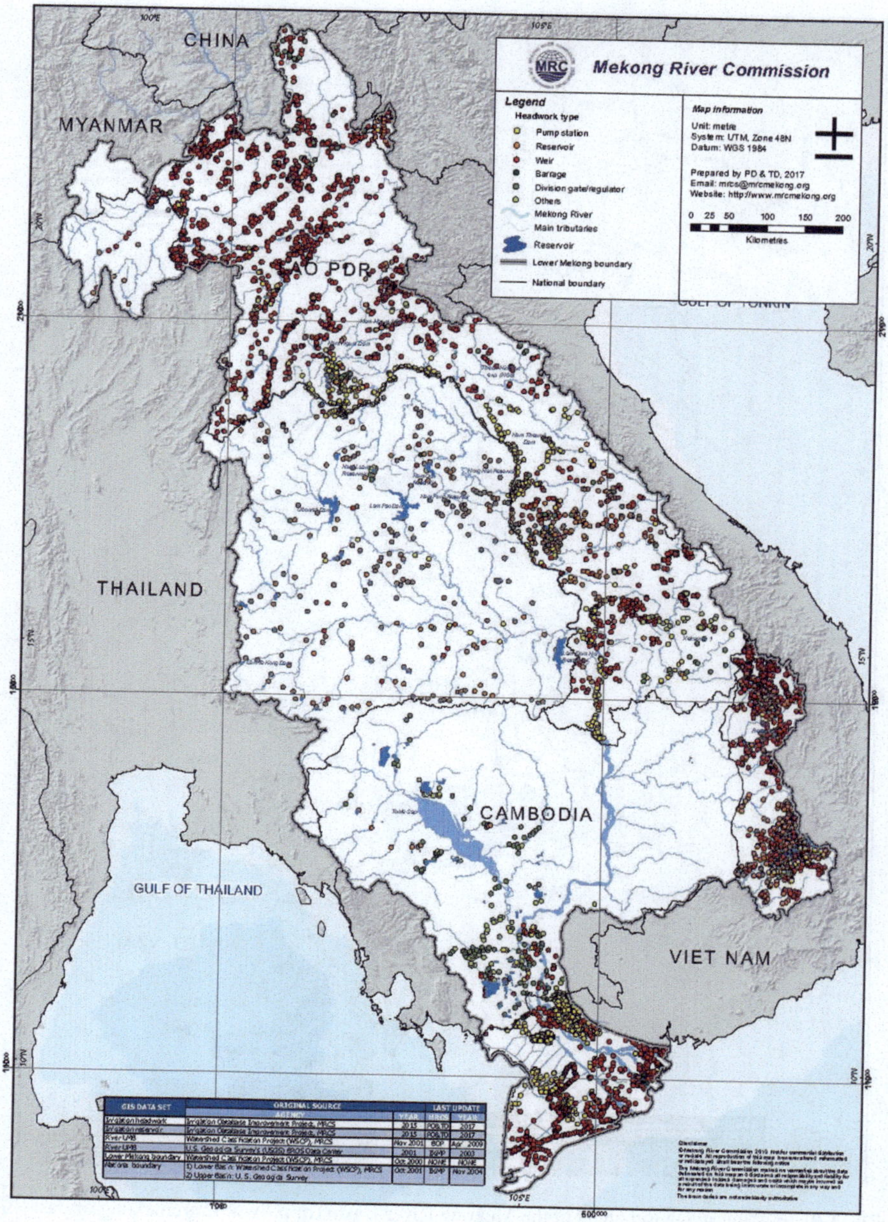

Fig. 1.8 Existing irrigation headworks and reservoirs in the Mekong River Basin (http://www.mrcmekong.org/assets/Publications/MRC-Irrigation-Database-Improvement-final-for-web.pdf, p 54)

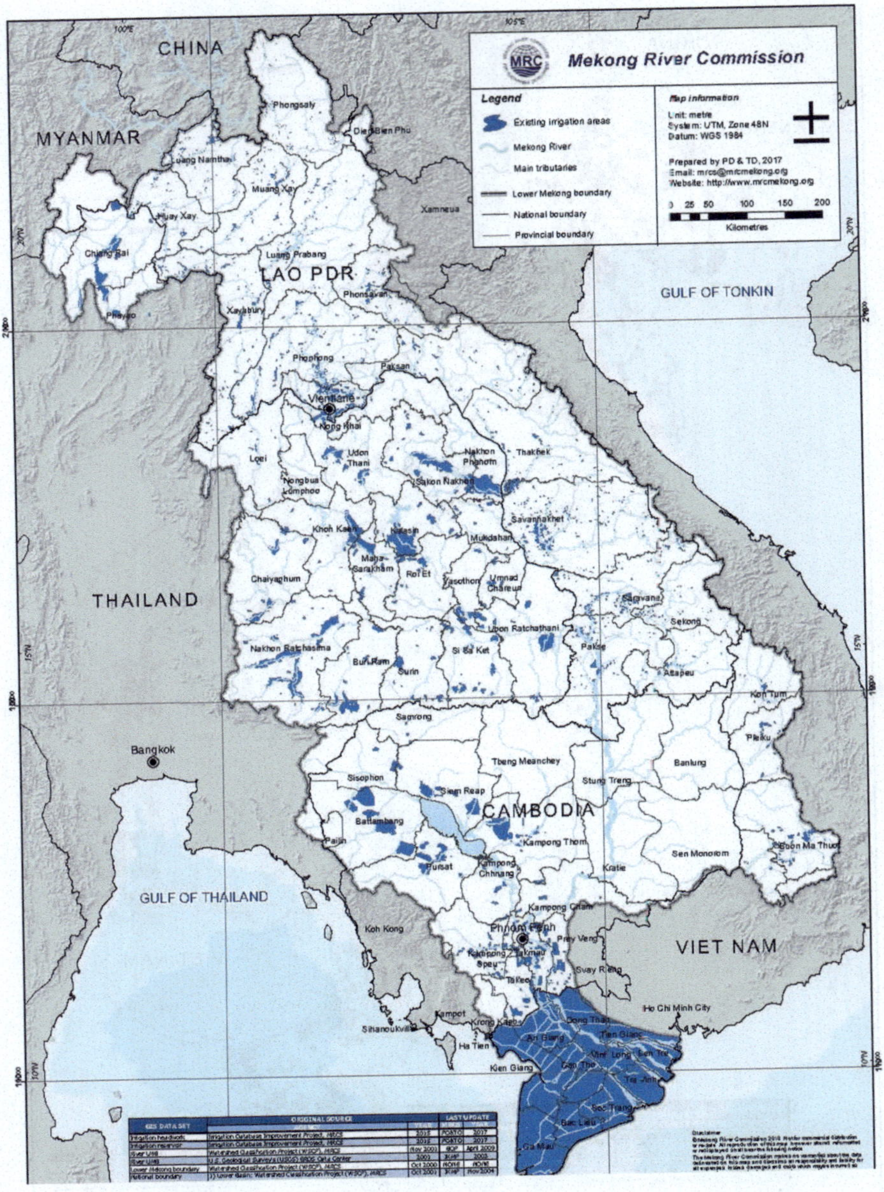

Fig. 1.9 Existing irrigation areas in the Mekong River Basin (http://www.mrcmekong.org/assets/Publications/MRC-Irrigation-Database-Improvement-final-for-web.pdf, p 55)

Table 1.7 Statistics of hydropower projects under construction or planned on the main stream and tributaries of the Mekong River [According to *Mekong R C. Planning Atlas of the Lower Mekong River Basin. Phnom Penh: Lao Uniprint Press Co. Ltd., 2011, p 79* and updated information of Lower Se San 2 project in Cambodia (http://www.hydrosesan2.com/project.php?id=119), Xayabury and Don Sahong projects in Lao PDR by 2016]

Country	Main stream		Tributaries			Total
	Under construction	Planned	Existing	Under construction	Planned	
Cambodia		2	2		10	14
Lao PDR	2	7[a]	11	9	71	100
Myanmar (Data of Myanmar is not available)	–	–	–	–	–	–
Thailand			7			7
Viet Nam			7	5	3	15
Total	2	9	26	14	85	136

[a]Two of them located on the transboundary between Lao PDR and Thailand on the mainstream Mekong River

constructed on the tributaries in the north, no more hydropower station is planned in following years. In Viet Nam, there are 7 hydropower stations existing, 5 under construction and 3 planned, which are all located on the tributaries. In specific, hydropower stations under construction or planned are situated in the Se San-Sre Pok Basin. The distribution of hydropower stations in the basin is demonstrated in Fig. 1.10.

The gross installed capacity of the hydropower projects existing, under construction and planned in the Mekong River Basin is 29,684 MW. According to statistic in 2011, existing hydropower stations register a gross installed capacity of 2688 MW, hydropower stations under construction 5320 MW and planned hydropower stations 21,676 MW, as shown in Table 1.8.

Hydropower projects with large installed capacity are mainly located on the mainstream, with an installed capacity of 14,697 MW or 49.5% of the gross installed capacity. In specific, hydropower projects on mainstream Mekong River in Lao PDR register an installed capacity of 10,417 MW or 70.9% of gross installed capacity of mainstream. The gross installed capacity of hydropower projects existing, under construction or planned on tributaries is 14,987 MW, mainly located in Lao PDR.

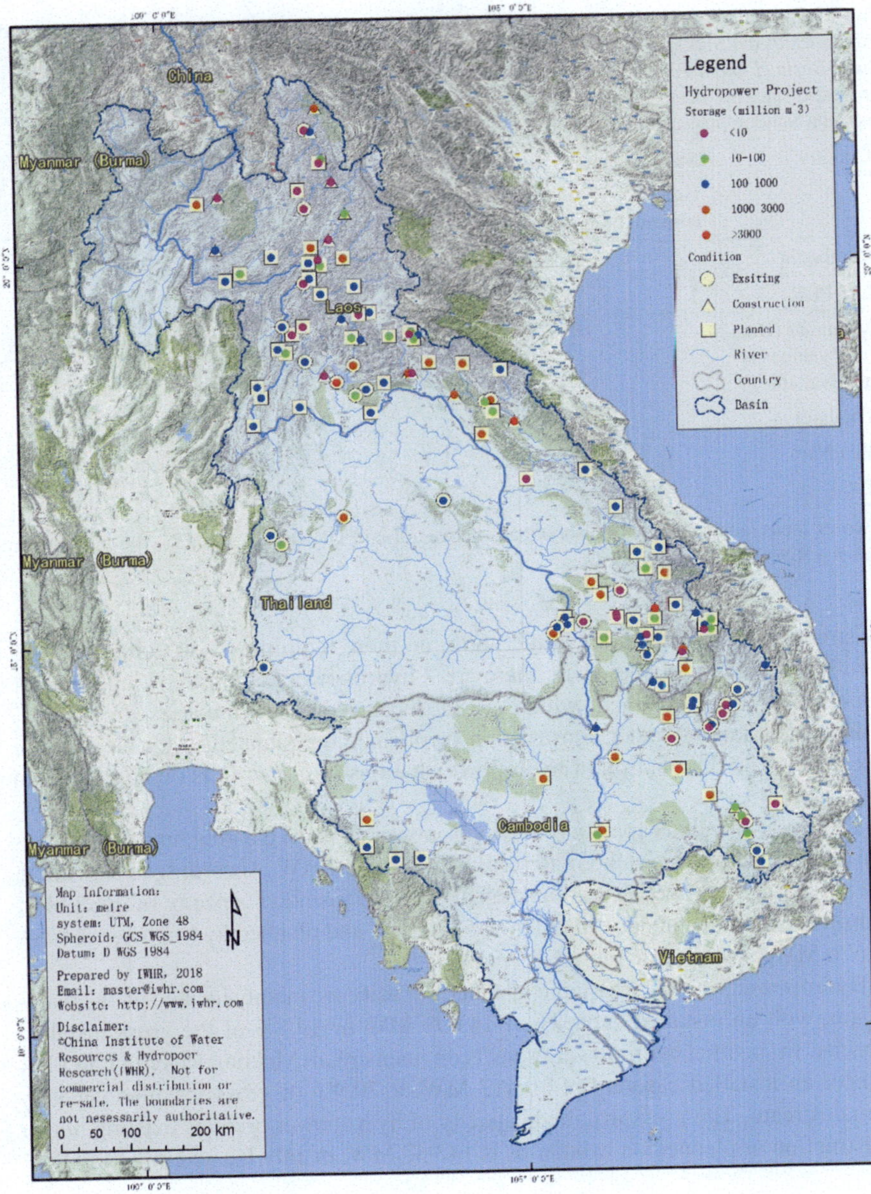

Fig. 1.10 Hydropower projects in the Mekong River Basin

Table 1.8 Statistics of gross installed capacity of hydropower in the Mekong River Basin [Mekong R C. Planning Atlas of the Lower Mekong River Basin. Phnom Penh: Lao Uniprint Press Co. Ltd., 2011, p 81 (The table was updated by the author according to hydropower constructions in Cambodia and Lao PDR by 2016)]. Unit: MW

Country	Main stream			Tributaries				Total
	Under construction	Planned	Total	Existing	Under construction	Planned	Total	
Cambodia		4280	4280	1		1309	1310	5590
Lao PDR	1540	8877[a]	10417	738	2764	6847	10,350	20,767
Thailand				745			745	745
Viet Nam				1204	1016	363	2583	2583
Total	1540	13157	14697	2688	3780	8519	14,987	29,684

[a]3879 MW installed capacity is located on mainstream Mekong River on the border between Lao PDR and Thailand

Dr. Baiyinbaoligao's research primarily focuses on Eco-hydraulics, urban river, and river restoration. He was the Program Manager of important national and provincial level projects. The number of investigated or participated projects and programs exceeds 70. He has written 4 technology standards (or specifications) and 4 books, has published over 80 papers.

Dr. Liu's research primarily focuses on hydrological processes and modeling, flood management, and transboundary river water cooperation. She has over 5 years' research experiences in transboundary water cooperation of China. She was a participant of various Lancang-Mekong water resources cooperation activities. She is an editor of a joint publication by MWR of China and MRC in 2016. She is Program Managers of 3 international cooperation projects with Mekong River countries. She is author of over 10 journal papers and 1 book.

Dr. Chen's research primarily focuses on comprehensive management of urban rivers, river ecological restoration, river sediment management, transboundary river water management and flood management. She was the Program Manager of many important national and provincial level projects. She has over 2 years' research experiences in transboundary water cooperation of China. The number of participated projects and programs exceeds 50. She has written 4 technology standards (or specifications) and 4 books, has published over 50 papers.

Dr. Mu's research primarily focuses on water resources management, river ecosystem restoration and hydraulic computation. He has over 10 years' research experiences in relative national level projects. He is author of over 40 journal papers and 4 books, as well as dozens of patents for inventions.

Chapter 2
Summary of Flood and Drought in Mekong River Basin

Xingru Chen, Hui Liu, Baiyinbaoligao, and Xiangpeng Mu

Abstract Flood and drought disasters occur frequently in Mekong River basin owing to ocean climate and climate change. Based on the collected floods and droughts in Mekong River basin in recent 20 years, the losses, causes and effects of flood and drought disasters are analyzed. The main conclusions are drawn as follows: (1) Flood-caused fatalities in the Mekong River Basin were most serious, 825, in 2000, followed by 2001 (489), 2011 (396), 2013 (247) and 1996 (173). Cambodia and the Cuu Long Delta, Viet Nam take the largest shares while the share is small in northeastern Thailand and Lao PDR. (2) Flood imposes significant impact on agriculture. 2000 saw the biggest agricultural impact of floods, approximately 2.50 million hm^2, followed by 2011, about 500,000 hm^2. Seen from the geographic distribution, agricultural impact gathered in the Cuu Long River Delta, Viet Nam in 2000, and Cambodia. (3) Flood incurs serious economic loss in countries in the basin. Cambodia ranks top by 1.4 billion USD, followed by Viet Nam by 980 million USD and Lao PDR by 590 million USD. Thailand ranks bottom by 310 million USD in 1996–2014. (4) Extreme

Myanmar's territorial area in the Mekong River Basin takes up approximately 3% of the drainage area of the Mekong River, and the mainstream Mekong River is the boundary river between Myanmar and Lao PDR. According to the statistical analysis of flood disaster data of Mekong River Basin countries, Myanmar's flood disasters mainly took place outside the Mekong River Basin. Therefore, emphasis is placed on the Mekong River Basin in Cambodia, Lao PDR, Thailand and Viet Nam when it comes to the analysis of flood disasters in the Mekong River Basin in this paper.

X. Chen (✉) · H. Liu · Baiyinbaoligao · X. Mu
China Institute of Water Resources and Hydropower Research, A-1 Fuxing Road, Haidian District, Beijing 100038, China
e-mail: chenxr@iwhr.com

H. Liu
e-mail: wuyouliuhui@163.com

Baiyinbaoligao
e-mail: baiyin@iwhr.com

X. Mu
e-mail: swood2002@163.com

H. Liu (ed.), *Flood Prevention and Drought Relief in Mekong River Basin*, Springer Tracts in Civil Engineering,
https://doi.org/10.1007/978-981-15-2006-8_2

floods take place more frequently and cause a huge loss upon the beginning of the twenty-first century. The flood loss amounted to 1.164 billion USD in 2011, taking up 35% of the gross flood loss in 1996–2014; to 651 million USD in 2013, taking up 20%; and to 462 million USD in 2000, taking up 14%. (5) Drought features extensive influence, long duration and huge economic loss. Seen from spatial distribution, drought loss is heavy in northeastern Thailand, Cambodia and Viet Nam and relatively mild in Lao PDR.

2.1 Objective

In order to effectively cope with flood and drought disasters in the basin in the future, it is necessary to deepen the recognition of the current status of flood and drought disasters in the Mekong River Basin, which will also lay foundation for the further analysis and modelling.

Records of flood and drought was collected and summarized, with a statistic of the loss in the past decades, which will provide a brief introduction of the spatial and temporal distribution of disasters in the basin.

Typical flood and drought were illustrated with detailed description based on data available, in order to provide the readers a better understanding of the whole picture of the typical events.

2.2 Data and Methodology

2.2.1 Data

Data of natural disasters of the Mekong River Basin was collected and analyzed. The data sources include:

- International Disaster Database (EM-DAT) established by the Center for Research on the Epidemiology of Disasters (CRED) of Belgium's University of Leuven;
- Asian Disaster Reduction Center (ADRC);
- ReliefWeb—a disaster statistics website sponsored by the UN Office for the Coordination of Humanitarian Affairs (OCHA) (https://reliefweb.int);
- Disaster data from the Mekong River Commission, including Annual Mekong Flood Report etc.

The EM-DAT database contains detailed recording of the occurrence time, place, type, fatalities, affected population, economic loss etc. Those data are from a variety of sources, including relevant UN organizations, NGOs, insurance companies, research institutions and publishing agencies. The database is a free shared database that records the most and complete information of disasters, is very reliable and authoritative, and has been widely quoted by many international institutions, organizations and scholars.

A disaster won't be included into EM-DAT until it meets one of the following three conditions: 10 or more people reported killed; 100 or more people reported affected; declaration of a state of emergency or request for international aid. Therefore, seen from the completeness and comprehensiveness of disaster recordings for years, international public data are inferior to national databases. But these data can meet the requirements in the macroscopic analysis of the spatio-temporal distribution of flood loss, especially as basis for relatively major disaster loss.

Every database has its characteristics. The EM-DAT database has a small number of parameters about single disaster event, including fatalities, affected population, occurrence place and economic loss parameters, but it is advantageous in the recording length and can analyze the long-sequence variation trend of a given disaster, but doesn't present detailed description of disaster events. The advantage of the disaster statistics web of the Asian Disaster Reduction Center (ADRC) lies in the relatively detailed description of single disasters, but the disadvantage is the short length of record and its data mainly start from the 1990s.

Disaster analysis of this assessment report is based on data from said various international public authoritative webs. Those data are verified and supplemented mutually, so the report is exempted from the problem of one-sided data that may be caused by taking a single database as the data source, and the statistical analysis and research of disaster loss in this project is based on solid fundamental data support.

2.2.2 *Methodology*

Based on the above-mentioned public data sources, we analyzed the spatio-temporal distribution characteristics of the occurrence time, occurrence place, fatalities, affected population, housing damage, affected farmland, traffic, school and economic losses and other parameters of flood and drought disasters in the countries located in the Mekong River Basin.

For the description of typical floods, the meteorological conditions, water level, discharge and flood loss were analyzed. The findings from previous publications were primarily used in this section.

For the description of typical droughts, some drought indicators (SPI and SRI) were applied. The descripition of these indexes please refer to that in Chap. 4.

2.3 Summary of Flood

2.3.1 *Flood Loss*

As a major disaster in the Mekong River Basin, flood threatens the life and property safety of people there by leading to housing damage, road closure and school suspension, affecting the order of production and life and causing damage or reduction of output in the most important industry—agriculture in the basin to a varying degree. In the following, fatalities, affected population, affected agricultural area and economic loss of flood disasters in the Mekong River Basin are analyzed respectively.

(1) **Flood-caused fatalities**

Table 2.1 showcases the fatalities caused by flood disasters in the Mekong River Basin in the recent 20 years. Fatalities and a huge life and property loss were witnessed in almost every flood disaster in the Mekong River Basin.

Figure 2.1 showcases the variation of yearly flood-caused fatalities in the basin. 2000 saw the most flood-caused fatalities, 825, followed by 2001 (489), 2011 (396), 2013 (247) and 1996 (173). The death toll is smaller than 100 in other years.

Figure 2.2 showcases the share of fatalities at each country in the basin. Cambodia and the Cuu Long River Delta, Viet Nam rank top. In specific, the biggest death toll

Table 2.1 The spatio-temporal distribution of flood-caused fatalities in the Mekong River Basin (Unit: Person)

Year	Cambodia	Lao PDR	Northeast Thailand	Cuu Long River Delta	Eastern Highlands of Viet Nam	Total
1996	169	–	–	–	4	173
2000	347	–	25	453	>20	825
2001	62	–	34	393	–	489
2002	–	3	–	71	2	76
2003	–	–	–	23	6	29
2004	–	–	–	38	–	38
2005	4	5	0	44	–	53
2006	11	5	–	55	0	71
2007	10	2	–	30	29	71
2008	–	7	–	7	7	21
2010	8	7	–	–	4	19
2011	250	42	–	89	15	396
2012	26	5	–	0	0	31
2013	168	17	17	45		247
2014	49	5	4	17		75

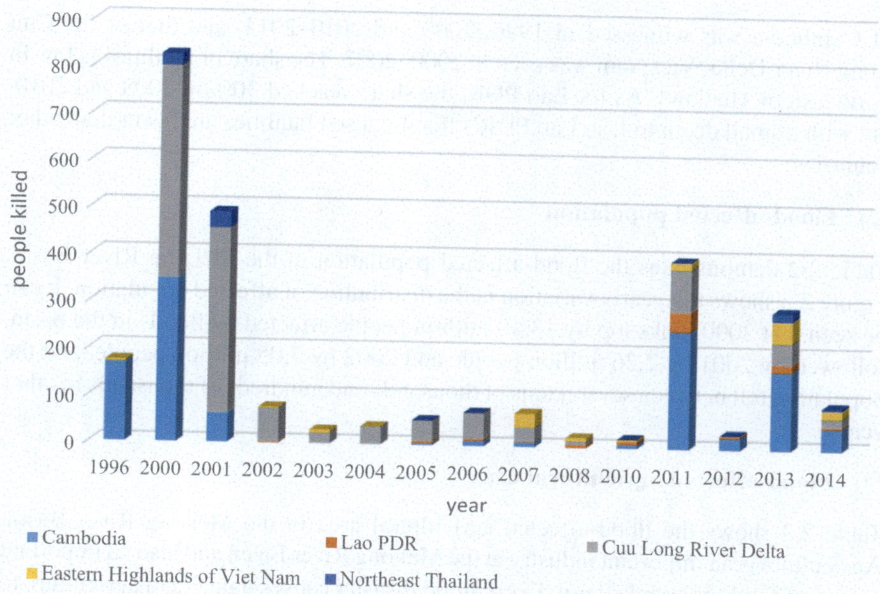

Fig. 2.1 Variation of yearly flood-caused fatalities in the basin

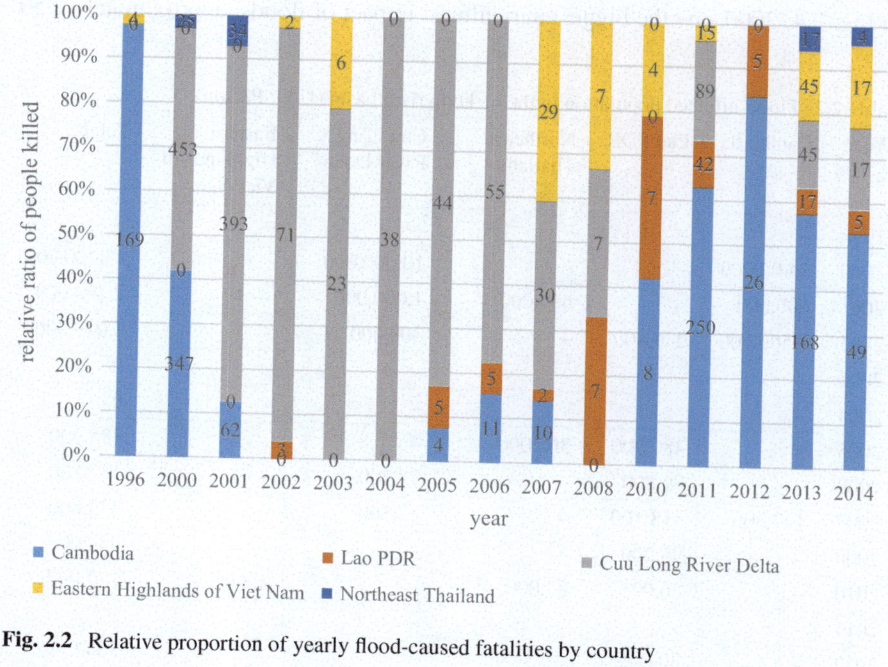

Fig. 2.2 Relative proportion of yearly flood-caused fatalities by country

of Cambodia was witnessed in 1996, 2000 and 2010–2014, and that of the Cuu Long River Delta, Viet Nam was seen in 2000–2007. The share of fatalities is low in northeastern Thailand. As for Lao PDR, the share reached 30% in 2008 and 2010, but with a small death toll, so Lao PDR's flood-caused fatalities are fewer than other countries.

(2) **Flood-affected population**

Table 2.2 demonstrates the flood-affected population in the Mekong River Basin. Figure 2.3 shows the yearly variation in the distribution of affected population. It can be seen that 2000 ranks top by 13.40 million people affected by floods in the basin, followed by 2001 by 2.26 million people and 2002 by 2.05 million people, and the population fell between several tens of thousands and hundreds of thousands in other years.

(3) **Flood-affected agricultural area**

Table 2.3 shows the flood-affected agricultural area in the Mekong River Basin. Agriculture is an important industry at the Mekong River Basin and also an important source of food for people there. Every flood disaster causes damage to agriculture to a varying degree, in addition to human life loss in the basin.

Yearly flood-affected agricultural area at the Mekong River Basin is showcased in Fig. 2.4. 2000 saw the biggest agricultural impact of floods, approximately 2.50

Table 2.2 Flood-affected population in the mekong river basin (Unit: Person)

Year	Cambodia	Lao PDR	Northeast Thailand	Cuu Long River Delta	Eastern Highlands of Viet Nam	Total
1996	–	–	–	–	–	–
2000	3,400,000	–	–	10,000,000	–	13,400,000
2001	600,000	–	660,000	1,000,000	–	2,260,000
2002	1,500,000	249,800	–	300,000	–	2,049,800
2003	–	–	–	–	–	0
2004	–	–	–	–	–	0
2005	–	480,900	305,000	–	–	785,900
2006	–	89,800	–	77,700	–	167,500
2007	147,200	118,100	–	67,500	–	332,800
2008	–	95,200	–	–	–	95,200
2010	–	86,097	82,000	–	4000	172,097
2011	–	–	–	–	–	
2012	–	3047	–	–	–	3047
2013	–	–	429,000	–	–	429,000
2014	–	92,165	400,000	800	–	492,965

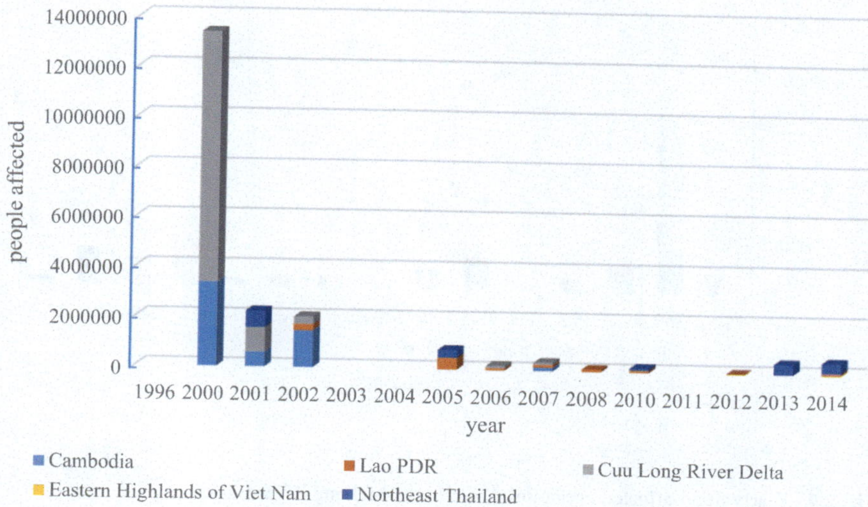

Fig. 2.3 Variation of yearly flood-affected population in the Mekong River Basin

Table 2.3 Flood-affected agricultural area at the Mekong River Basin (Unit: hm^2)

Year	Cambodia	Lao PDR	Northeast Thailand	Cuu Long River Delta	Eastern Highlands of Viet Nam	Total
1996	25,020	67,500	–	–	–	92,520
2000	421,600	42,900	–	2,000,000	–	2,464,500
2001	164,200	42,200	–	–	–	206,400
2002	45,000	33,700	–	–	9000	87,700
2003	–	800	–	–	1000	1800
2004	247,400	14,400	–	–	–	261,800
2005	55,000	56,000	39,500	–	–	150,500
2006	14,500	6900	–	14,700	130	36,230
2007	9500	7500	–	46,400	20,300	83,700
2008	18,900	28,500	–	28,500	80	75,980
2010	18,527	3861.1	14,000	–	5814	42,202
2011	284,300	77,000	–	250,000	–	611,300
2012	–	147,800	–	65,906	–	213,706
2013	–	–	301,000	–	–	301,000
2014	–	–	34,000	3098	–	37,098

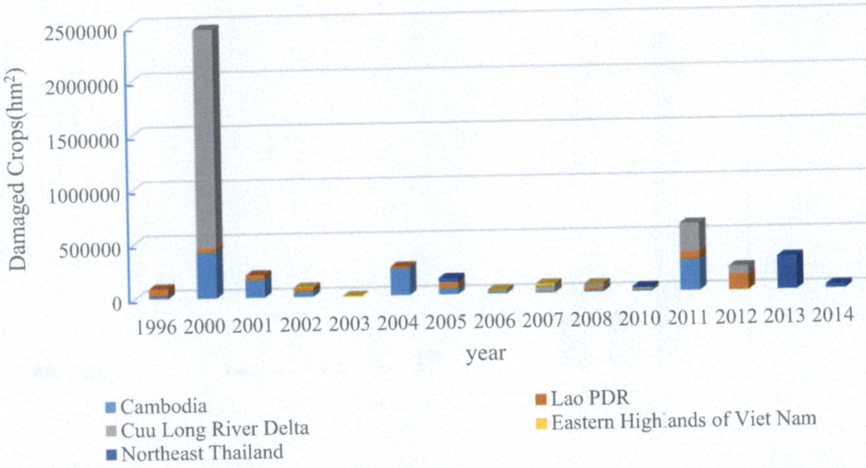

Fig. 2.4 Yearly flood-affected agricultural area at the Mekong River Basin

million hm^2, followed by 2011, about 500,000 hm^2. Seen from the geographic distribution, agricultural impact gathered in the Cuu Long River Delta, Viet Nam in 2000, and gathered mostly in Cambodia and the Cuu Long River Delta, Viet Nam and partially in Lao PDR in 2011. In 2013, floods mainly affected agriculture in the irrigation regions of northeastern Thailand. In 2001 and 2004, affected agricultural area was about 200,000–250,000 hm^2, largely located in Cambodia.

(4) **Flood-caused economic loss**

Floods cause reduction or damage of agricultural output, road closure, damage of water conservancy facilities, school suspension and housing damage, and yearly flood-caused economic loss at countries in the basin is summarized in Table 2.4. It can be seen that Cambodia and Viet Nam rank top in terms of the share of flood-caused economic loss by 43% and 29.6% respectively, followed by Lao PDR 17.9% and northeastern Thailand 9.6%.

Though all located in the Mekong River Basin, Cambodia, Lao PDR, Thailand and Viet Nam differ from one another in the temporal and spatial distribution of flood disasters because they have different hydrological and geographical conditions. For a detailed analysis on the inter-annual variation of flood loss in these countries, the variation in the yearly flood-caused economic loss of each country is expounded below.

Figure 2.5 shows the year-on-year flood-caused economic loss of Cambodia in the Mekong River Basin between 1996 and 2014. It can be seen that the country's most serious economic loss was seen in 2011 (624 million USD), 2013 (356 million USD) and 2000 (161 million USD).

Figure 2.6 shows the year-on-year flood-caused economic loss of Lao PDR in the Mekong River Basin between 1996 and 2014. The country's flood-caused economic

Table 2.4 Flood-caused economic loss at the Mekong River Basin (Unit: Mn USD)

Year	Cambodia	Lao PDR	Northeast Thailand	Cuu Long River Delta	Eastern Highlands of Viet Nam	Total
1996	86.5	10.4	–	113	–	209.9
2000	161	30	21	250	–	462
2001	36	56	23.9	99	–	214.9
2002	12.5	61	–	0.3	3	76.8
2003	–	18.3	–	15	0.5	33.8
2004	55	4.1	–	3	–	62.1
2005	3.8	18.3	2.8	3.5	–	28.4
2006	11.8	3.1	6.8	15	–	36.7
2007	9	18	–	1.5	50.8	79.3
2008	5.8	56	–	–	1	62.8
2010	62	21	47	55	–	185
2011	624	220	–	260	60	1164
2012	–	1.5	–	16	1	18.5
2013	356	62	210	23	0.2	651.2
2014	–	12	6	2.7	5.7	26.4
Total loss	1423.4	591.7	317.5	857	122.2	3311.8
Average loss	118.6	39.4	45.4	61.2	17.5	282.1
Percentage (%)	43.0	17.9	9.6	25.9	3.7	100

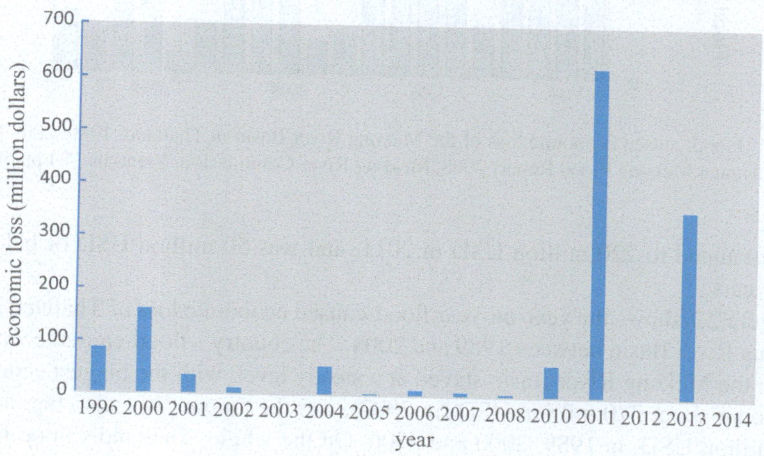

Fig. 2.5 Flood-caused economic loss in Cambodia, 1966–2014

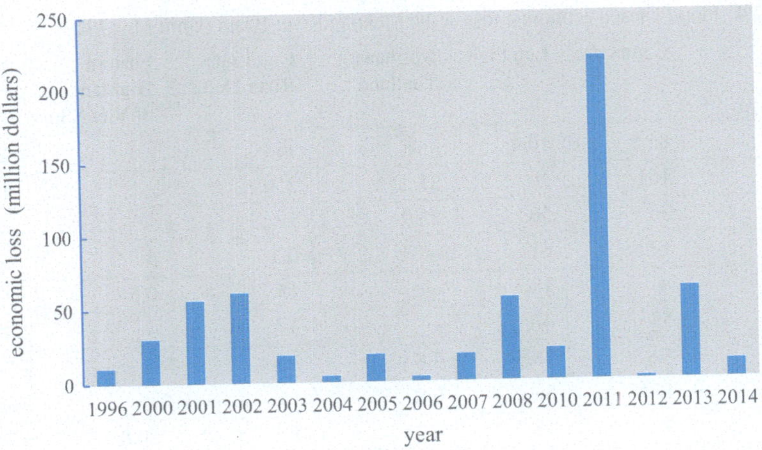

Fig. 2.6 Flood-caused economic loss in Lao PDR, 1966–2014

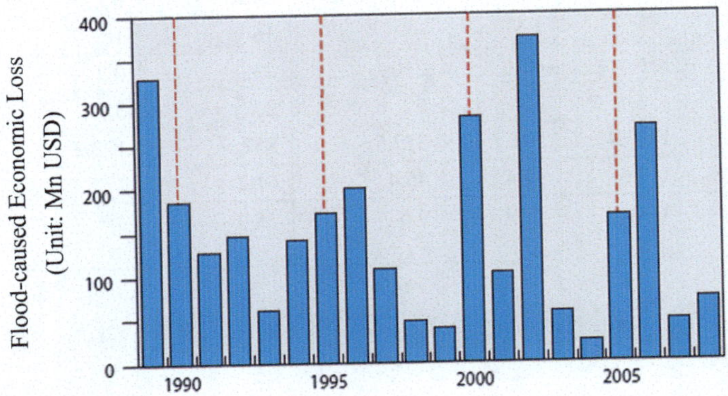

Fig. 2.7 Flood-caused economic loss of the Mekong River Basin in Thailand, 1989–2008 [MRC (2009) Annual Mekong Flood Report 2008, Mekong River Commission, Vientiane. 84 pp, p 27]

loss amounted to 220 million USD in 2011, and was 50 million USD or below in other years.

Figure 2.7 shows the year-on-year flood-caused economic loss of Thailand in the Mekong River Basin between 1989 and 2008. The country's flood-caused economic loss in the Mekong River Basin stayed at a steady level, with the biggest economic loss being about 370 million USD (in 2002), and the amount was also big, around 300 million USD, in 1989, 2000 and 2006. On the whole, Thailand's flood loss in the Mekong River Basin tends to fluctuate cyclically. Its economic loss caused by extreme floods in the Mekong River Basin in 2011 and 2013 remains unknown, which should be even hundreds of millions of USD according to Lao PDR's flood loss estimate.

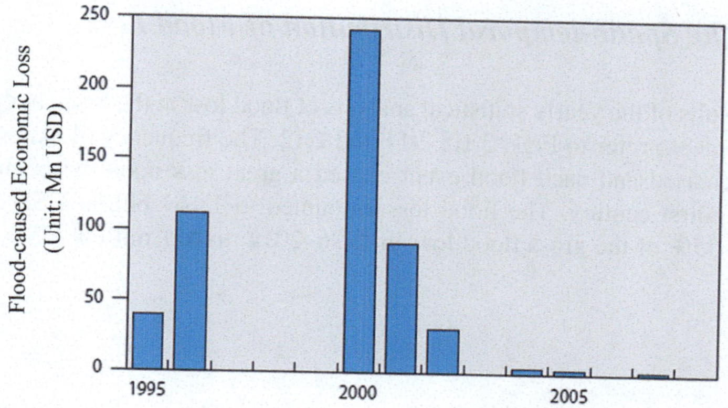

Fig. 2.8 Flood-caused economic loss of Mekong Delta in Viet Nam, 1995–2008 (Unit: Mn USD) [MRC (2009) Annual Mekong Flood Report 2008, Mekong River Commission, Vientiane. 84 pp, p 28]

Figure 2.8 shows the year-on-year flood-caused economic loss of the Mekong Delta in Viet Nam between 1995 and 2008. The most serious loss, 240 million USD, was seen in 2000, and the amount was about 100 million USD in 1996 and 2001.

Figure 2.9 shows the flood loss of the Mekong River Basin between 1996 and 2014. The flood loss in 2000, 2011 and 2013 takes up 69% of the gross flood loss. In specific, 2011 ranks top by taking up 35% of the 19-year flood loss, followed by 2013 by 20% and 2000 by 14%.

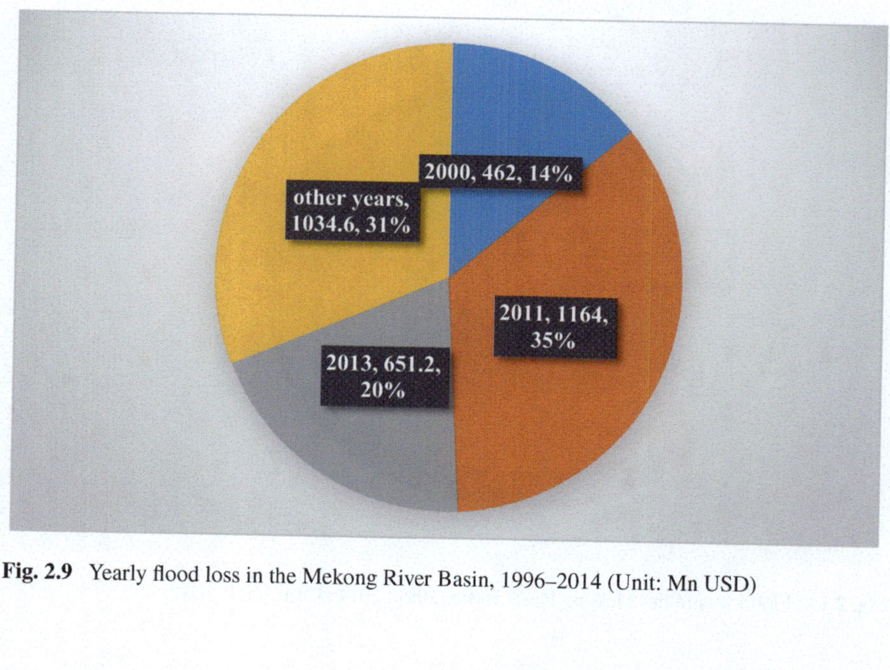

Fig. 2.9 Yearly flood loss in the Mekong River Basin, 1996–2014 (Unit: Mn USD)

2.3.2 *The Spatio-temporal Distribution of Flood Loss*

For the results of the yearly statistical analysis of flood loss at the basin in the recent 20 years, please refer to Figs. 2.10, 2.11 and 2.12. The frequency of extreme flood events increased and each flood event caused a great loss upon the beginning of the twenty-first century. The flood loss amounted to 1.164 billion USD in 2011, taking up 35% of the gross flood loss in 1996–2014; to 651 million USD in 2013,

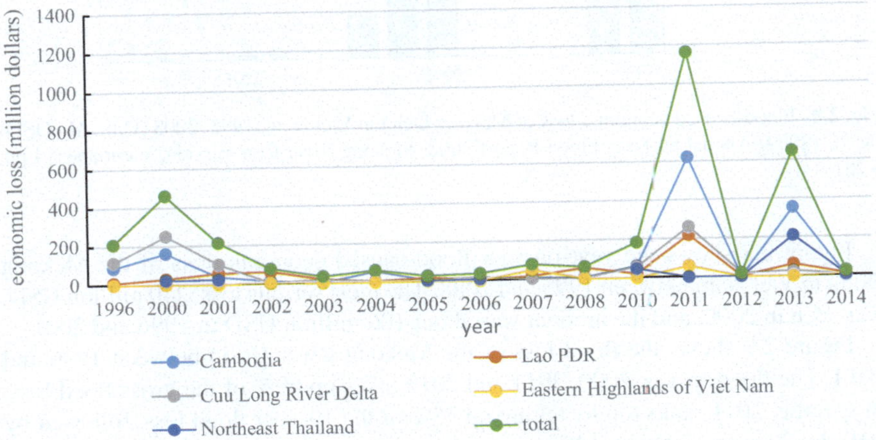

Fig. 2.10 Yearly flood loss of the Mekong River Basin by region

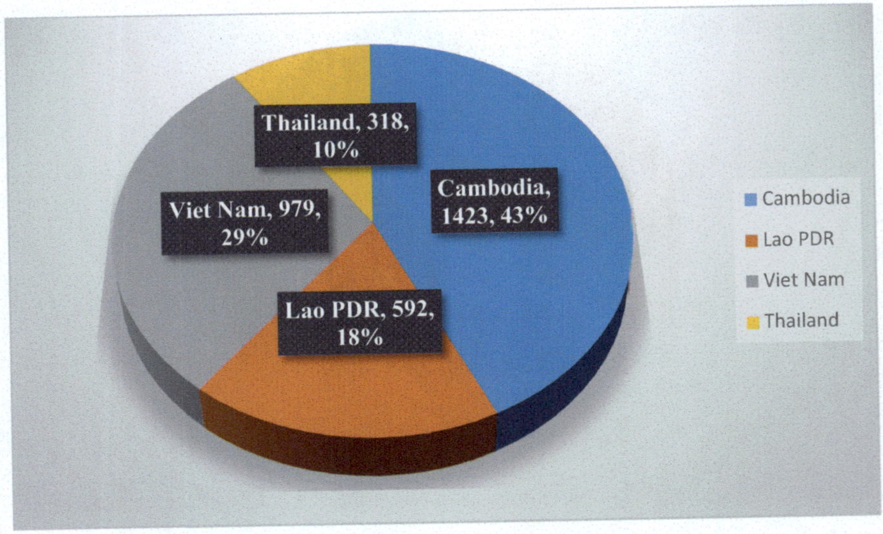

Fig. 2.11 Flood loss in the Mekong River Basin, 1996–2014 (Unit: Mn USD)

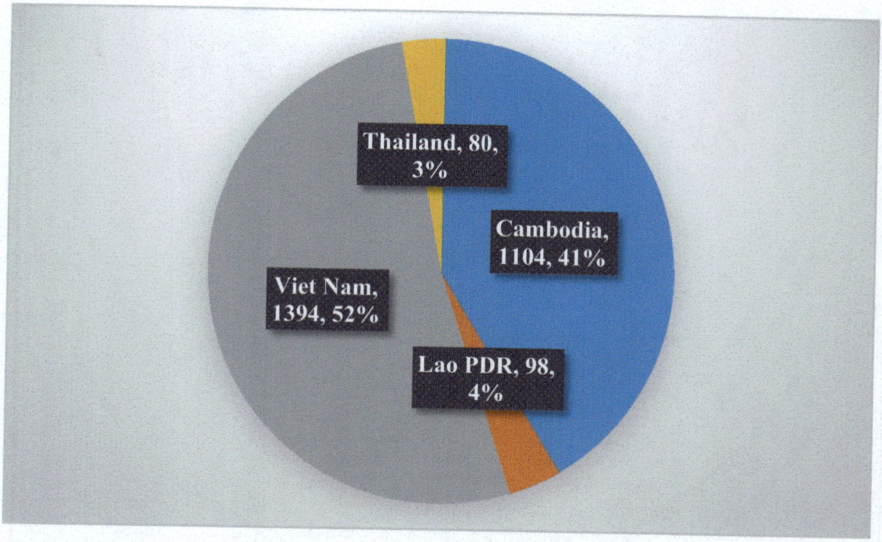

Fig. 2.12 Flood-caused fatalities in the Mekong River Basin, 1996–2014 (Unit: Person)

taking up 20%; and to 462 million USD in 2000, taking up 14%. The flood loss was heavy in those three years. Compared to small floods, economic losses caused by extreme floods in different regions at the basin in these countries are synergetic in time. Cambodia and Viet Nam register the largest flood loss and the largest share of fatalities.

Frequency statistics of flood disasters in the Mekong River basin was made from 1962 to 2017, the result is shown in Fig. 2.13. High flood risks are distributed in the lower reaches, especially in Cambodia and some provinces in central Lao PDR that close to Viet Nam. In the Mekong River Basin in Myanmar, Thailand and Viet Nam, flood risk is at an intermediate level. In provinces located in the upstream basin, flash floods are also caused by the oceanic climate, but at a low rate and with a limited area and a small population affected.

2.3.3 *Flood Benefit*

Flood is two-sided. It imposes negative impact on human production and life on the one hand and on the other hand, improves soil fertility by taking with it a great many organic matters. As for the Mekong River Basin, Cambodia's flood plains and Mekong Delta are among the earliest origins of Southeast Asian civilization, where people recognized early before 1800 that the millions of tons of sediment deposition rich in organic matters taken by floods could improve the agricultural productivity of the farmland in Cambodia's flood plains and Mekong Delta. Table 2.5 shows the

Fig. 2.13 Flood statistics of 1962–2017 in Mekong River Basin

Table 2.5 Gross benefit of Mekong River Floods for regional agricultural production [MRC (2009) Annual Mekong Flood Report 2008, Mekong River Commission, Vientiane. 84 pp, p 11]

Country	Agricultural value of flood (10^8 USD)	Remarks
Cambodia	10	2006 data. Basis of calculation: Paddy area takes up 90% of gross agricultural area, and 32% of paddy area is located in flood plains
Lao PDR	Non-significant	Agriculture is largely dependent on summer rain and autumn irrigation
Thailand	Non-significant	–
Viet Nam	35	2004 Mekong Delta data
Total	45	–

gross benefit the floods of the Mekong River bring to the agricultural production of Cambodia and Viet Nam.

Floods also bring significant benefits to the aquaculture industry represented by fishery, in addition to agriculture. When floods register long duration and wide coverage, the biologic amount of fishery will be high.

Flood is also of important significance for maintaining the ecological service value of wetland, which includes many functions such as food and freshwater supply, climate regulation, hydrological rhythm, pollution control, leisure and entertainment, aesthetics and education, biodiversity protection and nutrient cycling. Figure 2.14 shows wetland of Lao PDR along the Mekong River.

Fig. 2.14 Wetland in Lao PDR

2.4 Summary of Drought

Drought means the precipitation in a period is apparently smaller than multi-year average precipitation, leading to a relative shortage of water for a series of water circulation elements. The water circulation process follows an axis from atmospheric process (precipitation), surface process (surface water) to underground process (soil water, groundwater). Water deficit in a link will affect the next link, and these links will influence one another along with water circulation progress. Water deficit in any stage of water circulation will lead to drought in the corresponding form. When rainfall of a region is deficit compared to the historical level, the region will suffer meteorological drought; along with the progression of the drought, soil moisture will decline continuously because of the lack of rainfall replenishment, which will make it difficult for the effective replenishment of river runoff and thereby, will cause the hydrological drought that features slower and smaller flow. Based on the aforementioned formation mechanism, American Meteorological Society (AMS) classifies drought as: meteorological drought (precipitation decreases evidently compared to the same period of normal years), hydrological drought (long-term deficiency of precipitation leads to shortage of surface water and groundwater compared with multi-year average) and agricultural drought (water requirement for growth of crops is unsatisfied because of soil moisture deficiency). In specific, meteorological drought mainly affects rain-fed paddy, taking up around 90% of paddy in the Mekong River Basin in Lao PDR, Thailand and Cambodia, so agriculture of the Mekong River Basin is under great influence of drought.

2.4.1 *Drought Loss*

Drought loss in the Mekong River Basin is mainly made up of agricultural loss, such as reduction or devastation of yield, especially for paddy. In addition, fishery and livestock yields may be also reduced. According to MRC "Report of Flood and Drought Management and Mitigation Programme 2011–2015", EM-DAT database and ADRC data, the recent drought losses in the Mekong Basin are as in Table 2.6.

It should be noted the drought loss in said table is merely the estimated loss in a given aspect, so the drought loss is understated. Even so, the loss remains huge. The drought loss of the Cuu Long River Delta in 2002 and May 2004 is much higher than that of the loss of 7 out of the 9 floods in 2000–2008. In northeastern Thailand, drought-caused paddy loss is about 10 million USD/year, and its impact on aquaculture of the Tonle Sap Lake is also very apparent, with a yearly loss of 15 million USD/Year.[1]

According to the survey of drought loss of each country in the basin, droughts cause bigger affected population and economic loss than floods, though drought

[1] http://www.mrcmekong.org/assets/Publications/basin-reports/FMMP-working-paper-110820.pdf, p iii.

Table 2.6 Estimate of drought loss in the Mekong River Basin

Country	Year	Estimated loss (Mn USD)	Evaluation	Data source
Cambodia	1987	–	–	EM-DAT
	1994.6	100	–	EM-DAT
	1998.9	14.5	Fishing loss of the Tonle Sap Lake	WFP, 2009
	1999	6	Mekong Delta	MRC working paper (2011–2015)
	2001.9	–	–	EM-DAT
	2002.3	22	Mainly paddy output	IRRI, 2007
	2003.4	14.5	Fishing loss of the Tonle Sap Lake	MRC working paper (2011–2015)
	2004.5	21	Mainly paddy output	WFP, 2009
	2007.8	14.5	Fishing loss of the Tonle Sap Lake	MRC working paper (2011–2015)
	2016.5	–	Provinces near Tonle Sap Lake	EM-DAT
Lao PDR	1977	–	Southern provinces	EM-DAT
	1987	–	Northern provinces	
	1988.12	–	Southern provinces	
	1991.7	1	National	
	1999.1	–	–	
Myanmar	No record in the EM-dAT or ADRC database			
Northeastern Thailand	1991.3	–	–	EM-DAT
	1993	2	North and Middle Thailand	EM-DAT
	1999.1	–	–	EM-DAT
	2002.2	2.3	Northeastern Thailand (Mekong River Basin part)	EM-DAT

(continued)

Table 2.6 (continued)

Country	Year	Estimated loss (Mn USD)	Evaluation	Data source
	2005.1–2005.3	–	Thailand, including some provinces in the Mekong River Basin	EM-DAT
	2008.4	–	Most provinces of Thailand, including some provinces in the Mekong River Basin	EM-DAT
	2010.3	–	Most provinces of Thailand, including some provinces in the Mekong River Basin	EM-DAT
	2011.6	–	Most provinces of Thailand, including some provinces in the Mekong River Basin	EM-DAT
	2012.4–2012.8	1.2	Most provinces of Thailand	EM-DAT
	2014.3	–	42 provinces in northern Thailand, 28 provinces in the northeast	EM-DAT
	2016.5	–	Only paddy output, loss may be underestimated	EM-DAT
	Annual Loss	10	Thailand, including some provinces in the Mekong River Basin	MRC working paper (2011–2015)
Viet Nam Cuu Long River delta	1997.12–1998.4	–	National	EM-DAT
	1998.12–1999.4	6	Cuu Long River delta	EM-DAT
	2002	24	Mekong Delta	Mekong Flood Report

(continued)

Table 2.6 (continued)

Country	Year	Estimated loss (Mn USD)	Evaluation	Data source
	2004.7–2004.11	42	Mekong Delta	MARD, 2005
	2004.11–2005.4	110	Middle and South Viet Nam	EM-DAT
	2016.5	675	Mekong Delta	EM-DAT

doesn't take place frequently, and individual droughts feature extensive impact, long duration and huge economic loss and the gross loss incurred thereof tends to increase significantly.

Cambodia records 10 drought events, of which the 1987 drought isn't recorded with occurrence place, affected population and loss data. 6 of the other 9 droughts took place after 2000, indicating drought frequency increased significantly in the recent 20 years in Cambodia. Seen from affected population, the 1994 drought incurred a loss of 100 million USD, and the loss of other droughts fell between 6 million USD and 22 million USD.

Lao PDR encountered 5 droughts, mainly between 1987 and 1999, and no drought was reported since the beginning of the twenty-first century. In the country, a single drought imposes extensive influence in many provinces and even across the country. On the whole, droughts feature small frequency and extensive influence in Lao PDR.

In Myanmar, drought frequency is low, and the country has no drought report no matter on Belgium's EM-DAT web or Asian disaster reduction web.

In northeastern Thailand (the part in the Mekong River Basin), a total of 11 drought events took place, mainly after 2000, which indicates the country's drought frequency is also on a prominent rise upon the beginning of the twenty-first century. Seen from the economic loss of droughts, drought loss of 72 provinces in Thailand amounted to 3.3 billion USD in 2016 and 420 million USD in 2005. According to conservative estimate based on nationwide drought loss data, the drought loss in northeastern Thailand was 1.0–2.0 billion USD in 2016 and 100–200 million USD in 2005.

Drought recording of the Mekong Delta in Viet Nam started from 1997 to 2016, during which 6 drought disasters took place. Compared to other countries, its drought loss is the most serious though at a low frequency, and single drought's loss is about 24 million USD.

2.4.2 *Typical Drought Disasters*

Abnormally deficient rainfall is a critical factor driving the formation and occurrence of drought. When the precipitation in a region is small for a long time compared to the same periods in history, meteorological drought will be formed in the region; along with the further development of drought, soil moisture will start to decline because of the persistent shortage of rainfall replenishment, effective replenishment of river runoffs will be difficult and thereby, river discharge will decrease and hydrological drought will occur. Over the past few decades, countries along the Mekong River have experienced different degrees of drought events, causing tremendous impacts on agriculture, fisheries and production and life. This research selected two typical drought events in 2004–2005 and 2016 to reveal the causes, development process and impact of drought from the atmospheric circulation, meteorology, hydrology and other backgrounds. Specifically, the standardized precipitation index (SPI) (SPI1, SPI3, SPI6 and SPI12)[2] for diagnosing meteorological drought was established on different temporal scales of one month, three months, six months and twelve months based on the rainfall data of the past 116 years; and three typical hydrological stations, Chiang Saen, Mukdahman and Stung Treng, were selected as representatives of hydrological characteristics of upper, middle and lower reaches of the mainstream Mekong River, as well as the one-month, three-month, six-month and twelve-month standardized runoff indexes (SRI1, SRI3, SRI6 and SRI12) were calculated respectively based on each station's average monthly cross-section runoff sequence from January 1985 to December 2016. Based on the two established representative indicators of SPI and SRI, the causes and development of typical drought were analyzed.

1. **Drought in 2004–2005**

Since September 2004, most of the middle-east equator Pacific Ocean has been controlled by positive SST anomaly above 0.5 °C and the SST anomaly index in the NINO composite region (NINO1 + 2 + 3 + 4) has reached 0.6, indicating that the atmosphere-ocean in the tropical Pacific Ocean has entered the El Nino state; during the next 3–4 months (October 2004–January 2005), the sea surface temperature in the middle-east equator Pacific Ocean continued to be warmer and the most of the equatorial Pacific maintained a positive SST anomaly above 0.5 °C. Southern Oscillation Index continued to be negative and the El Nino phenomenon continued; until May 2005, the Pacific sea surface temperature returned to normal, marking the end of the El Nino event.[3]

Under the background of tropical Pacific ocean-atmosphere circulation anomaly, the early end of the wet season in 2004 in the Mekong River Basin led to widespread rainfall deficit in Thailand, Cambodia, Viet Nam and other countries. Figure 2.15 shows the spatial distribution of SPI6 in the Mekong River Basin in the dry season (November 2004–April 2005) and wet season (May 2005–October 2005) and SPI12 in the whole year (November 2004–October 2005). Figure 2.15 shows that in the dry season of 2004–2005, except for a small part of the northern part of the basin, the SPI of the other areas was generally smaller than −1, indicating that the precipitation throughout the basin was apparently smaller than former years and indicating the moderate or above meteorological drought. The SPI of the Thai-Lao border, central

[2]The definition of SPI and SRI please refer to Chap. 4.

[3]Wang Qiyi, Ren Fumin. The latest ENSO monitoring. Date: Mar. 16, 2005; news source: Climate System Diagnosis and Forecast Office. http://ncc.cma.gov.cn.

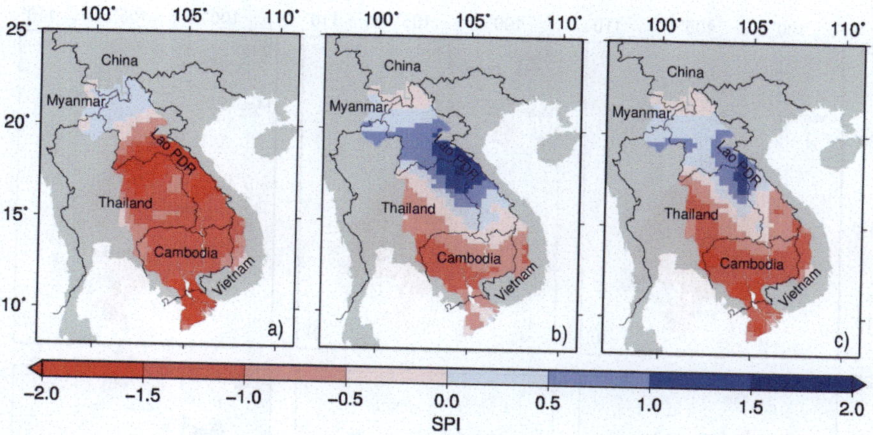

Fig. 2.15 Spatial distribution of SPI6 in the dry and rainy seasons and SPI12 in the whole year in 2004–2005 in the Mekong River Basin. **a** November 2004–April 2005; **b** May 2005–October 2005; **c** November 2004–October 2005

Lao PDR and the Mekong Delta Region was much smaller than −1.5, indicating that the drought in the above areas has reached the degree of exceptional drought. In the wet season of 2005, the drought coverage decreased significantly and the severity declined significantly compared with the dry season. The areas with less precipitation were mainly concentrated in northeastern Thailand, Cambodia and the Mekong Delta in Viet Nam and other middle and lower reaches. The SPI value of southwestern Cambodia was generally lower than −1, indicating the moderate drought in this area. Relatively speaking, countries such as Myanmar and Lao PDR in the upper reaches have abundant rainfall, high SPI and no drought. On the yearly scale, the drought in November 2004–October 2005 was mainly concentrated in the southern countries in the middle and lower reaches, including eastern Thailand, southern Lao PDR, Cambodia and the Mekong Delta in Viet Nam and so on; in terms of drought severity, the drought in the south was more severe than that in the north, reaching severe drought.

Figure 2.16 further shows the spatial distribution of monthly SPI3 in 2004–2005 dry season in the Mekong River Basin. In November 2004, SPI was calculated based on the cumulative rainfall of September–November (three months) in 2004, and was analogized in other months. It is known from Fig. 2.16 that in November 2004, the rainfall in the whole Mekong River Basin was lower than the historical level of the same period and the SPI value was negative, indicating that the whole basin was affected by drought. Among them, the drought in the southern region of the basin (northeastern Thailand, southern Lao PDR, Cambodia and Viet Nam and other countries) was particularly serious and SPI values were lower than −1.5, indicating severe drought; in December, the rainfall continued to be low and the drought in the Mekong River Basin continued to aggravate, as well as the scope of severe drought continued to expand; in 2005, the

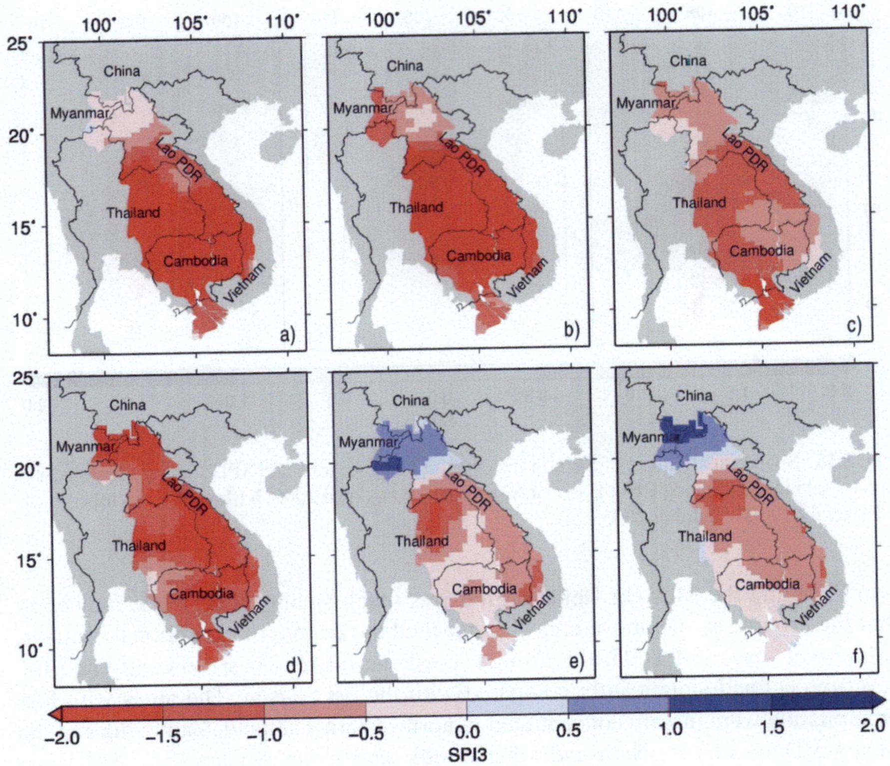

Fig. 2.16 Spatial distribution of monthly SPI3 in 2004–2005 dry Season in the Mekong River Basin (**a** November 2004; **b** December 2004; **c** January 2005; **d** February 2005; **e** March 2005; **f** April 2005)

whole basin in the first two months (January and February) was still in a continuous situation of less rain and drought was still severe; after entering the late dry season (March 2005–April 2005), the basin began to rain gradually and the SPI values of the northern parts of the basin (such as Myanmar, northern Lao PDR, etc.) returned to positive value and the regional drought subsided; although the southern region was still in a state of rainfall deficit, the overall drought has been effectively alleviated and the degree of drought has gradually been weakened to a mild level.

Figure 2.17 shows the six-month SRI (SRI6) sequences of three typical hydrological stations, Chiang Saen, Mukdahman and Stung Treng, in the upper, middle and lower reaches of the mainstream Mekong River. The SRI of Chiang Saen station in the upper reaches was lower than −0.5 in 2004–2005 and close to −2 in some periods, indicating that the runoff of this station in 2004–2005 was much lower than that of the same time in the previous years and the relatively severe hydrological drought occurred; compared with Chiang Saen station, the runoff of Mukdahman station was obviously lower in 2004–2005 dry season and the hydrological drought reached moderate or above level; compared with two stations in the upper and middle

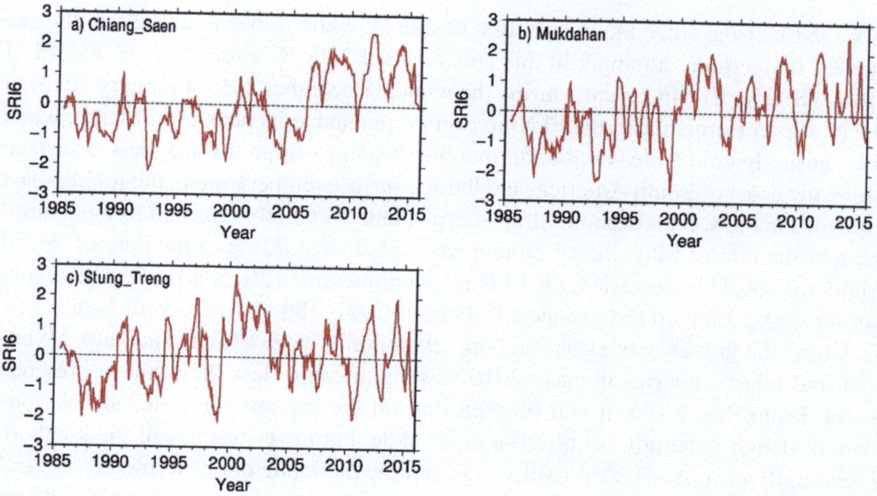

Fig. 2.17 SRI (SRI6) sequences of Chiang Saen, Mukdahan and Stung Treng stations

reaches, the dry season runoff of Stung Treng station in 2004–2005 did not show obvious anomalies, and the hydrological drought occurred mainly in the early months of 2005.

Influence of drought in 2004–2005

In 2004–2005, drought events imposed very serious impact on Viet Nam, Cambodia and Thailand in the basin.

In Cambodia, drought events that took place in 2004–2005 are the most serious in recent years, affecting 14 out of the 24 provinces, reducing paddy cultivation in all provinces and incurring a food shortage for 500,000 people.

Thailand suffered exceptionally serious droughts in 2004–2005, which affected 63 out of the 76 provinces. Across the country, 9 million people's life was affected, and irrigation water was restricted (and even prohibited) for assuring domestic water supply. Its gross loss amounted to 193 million USD, and data about the Ko Kret Plateau are absent.

The wet season of 2004 ended early, leading to the failure of autumn paddy cultivation in Viet Nam, especially in the Cuu Long Delta where more than 104,000 hm^2 of paddy was damaged. The impact was most serious in Ben Tre Province, where about 7000 hm^2 of paddy and 15,000 hm^2 of fruit farms were damaged, involving a value of 33 million USD. In addition, 82,000 households were forced to buy drinking water (4.5 USD/m^3). In the Cuu Long Delta, gross drought loss amounted to 42 million USD.

2. **Drought in 2016**

In August 2015, the sea surface temperature anomaly over Nino 3.4 in the tropical Pacific region (i.e. degree of deviation from the climate mean state) exceeded 2 °C (for the first time since the twenty-first Century), while in November 2015, the sea surface temperature anomaly in this area reached 2.95 °C, exceeding the record of 1997–98 super El Nino event. During the winter (December 2015–February 2016) in the Northern Hemisphere, the El Nino event remained at its peak until March, when the unusually cold water on the thermocline began to surge toward the sea surface along the coast of South America, and the sea surface temperature in the middle-east equator Pacific Ocean began to drop sharply, and eventually returned to the neutral state at the end of May, the El Nino event ended. According to the data of World Meteorological Organization, the El Nino phenomenon in 2015–2016 was relatively strong, comparable to the strongest El Nino in 1997–1998 since records began.[4]

Under the influence of strong El Nino, the rainfall in Viet Nam, Thailand, Myanmar and other countries in early 2016 was significantly less than that in previous years. From Fig. 2.18a, it can be seen that on the one-month scale, the obvious rainfall deficit generally occurred in the middle and lower reaches of the southern Mekong River in April 2016 (SPI < -1) and the moderate or above meteorological drought occurred. Especially in the southeastern pat of the basin, including southern Lao PDR, eastern Cambodia, Viet Nam and a small part of the eastern Thailand, their SPI values were lower than -1.5, indicating that the drought in these areas reached the exceptional drought level in the same month. When the temporal scale is extended to three months (Fig. 2.18b), it can be found that the drought coverage based on three-month cumulative rainfall monitoring (February–April 2016) was further extended to the entire Mekong Delta region, covering northeastern Thailand, central and southern Lao PDR, Cambodia and the Mekong Delta; at the same time,

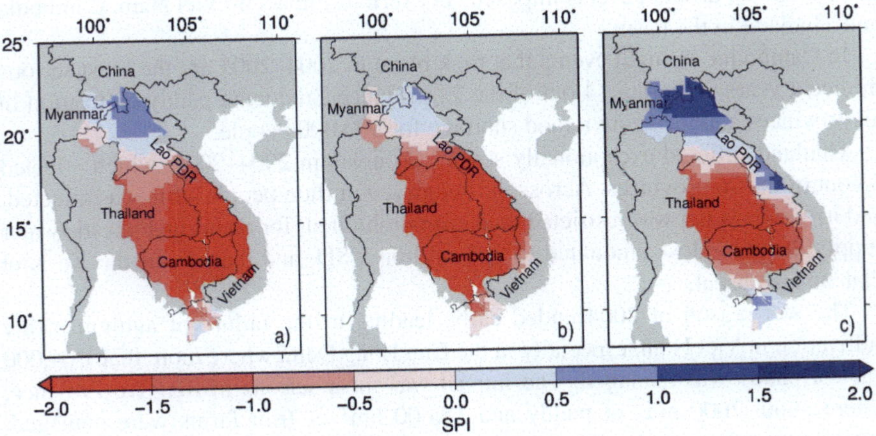

Fig. 2.18 SPI distribution in the Mekong River Basin on the one-month, three-month and six-month scales in 2016. **a** SPI1: April 2016; **b** SPI3: February–April 2016; **c** SPI6: November 2015–April 2016

[4]Mekong River Commission and Ministry of Water Resources of the People's Republic of China (2016). Technical Report – Joint Observation and Evaluation of the Emergency Water Supplement from China to the Mekong River. Mekong River Commission, Vientiane, Lao PDR, p 12.

the SPI value was generally lower than −1.5, indicating the severe and exceptional droughts; the diagnostic results on the six-month scale (November 2015–April 2016) show that the drought was mainly concentrated in northeastern Thailand and Cambodia and reached severe drought level. Compared with the diagnostic results on the three-month scale (Fig. 2.18b), there was no meteorological drought on the six-month scale (SPI > 0) in the Mekong Delta.

Figure 2.19 further shows the spatial distribution of monthly SPI3 in the Mekong River Basin from November 2015 to April 2016. It can be seen from Fig. that in November 2015, the Mekong River Basin was slightly dry and only a small range of mild drought took place in eastern Cambodia; in December 2015 and January 2016, most of the basin was wet; in February, the drought occurred, but mainly in the area with the border area of Thailand, Cambodia and Lao PDR as the center while most of the northern region was still wet; in March, the drought was widened to most parts of northern Thailand and Cambodia and generally reached the moderate level; as rainfall continued to be low, the drought in April was further expanded and covered all areas of the basin except northern Lao PDR and the severity of the drought reached severe level.

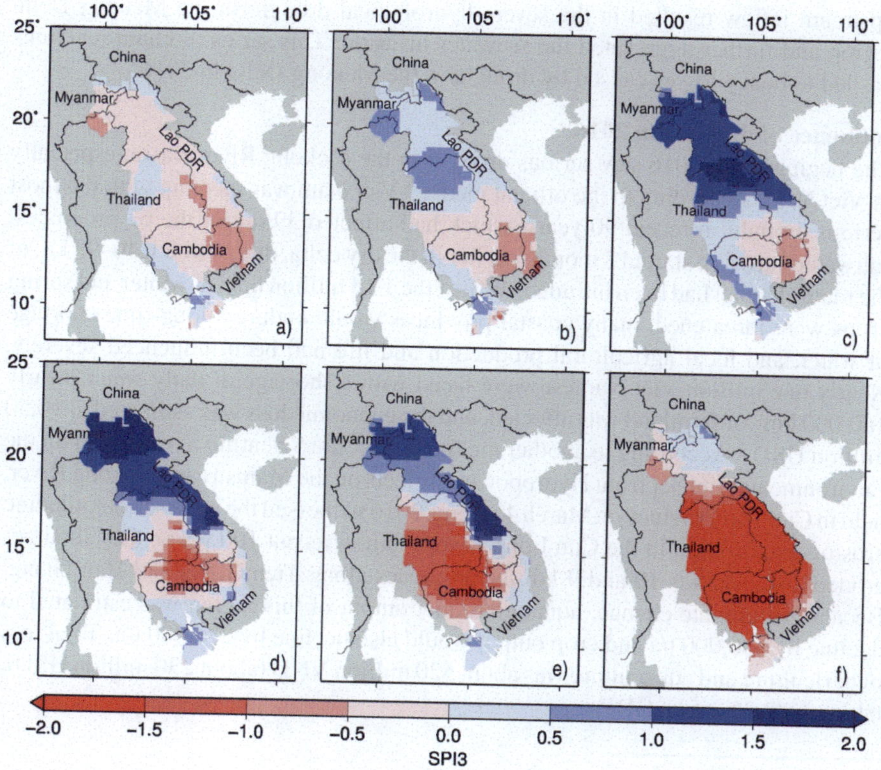

Fig. 2.19 Spatial distribution of monthly SPI3 in 2015–2016 dry season in the Mekong River Basin (a–f in Fig. is November 2015–April 2016 respectively)

Based on Fig. 2.17, the hydrological drought situation of three typical hydrological stations in 2016 was further analyzed. From the figure we can see that the SRI of Chiang Saen station in the upper reaches of the river in 2016 was basically above zero, indicating that the observed runoff of the station in 2016 was not lower than that of the same period in the previous years and there was no hydrological drought; for Mukdahan station in the middle reaches of the river, the SRI value in 2016 was lower than −0.5, indicating that the station had a certain degree of hydrological drought, but the drought was relatively mild. Compared with the two stations in the upper and middle reaches of the basin, the runoff of Stung Treng station in the lower reaches in 2016 was obviously lower than that in previous years and SRI value was even close to −2, showing a very serious hydrological drought. It can be seen that, compared with the upstream stations, the downstream stations experienced more severe hydrological droughts in 2016.

Combined with the above meteorological and hydrological analysis, February–April 2016 is the most extensive and influential period of the drought event. Among them, the sustained low rainfall in February–April triggered a certain degree of meteorological drought in the Mekong Delta region, but far less than that in Cambodia and Thailand in the upper reaches; however, the less rainfall and the reduction of upstream inflow resulted in the severe hydrological drought in the Mekong Delta region and further aggravated the seawater invasion. This series of chain reactions has led to heavy losses caused by drought in the Mekong Delta in 2016.

Influence of drought in 2016

The beginning of 2016 saw serious drought in the Mekong River Basin, especially in Viet Nam. According to the official media,[5] Viet Nam was meeting with the most serious drought in recent 90 years, which had affected 39 out of the 63 provinces; saltwater intrusion started 2 months earlier in the dry delta, due to which 40–93 km of the main channel had been intruded, 30% of the 1.15 million hm^2 of winter and spring crops were threatened, many coastal provinces would endure a long-time shortage of water, and local agricultural production and life had been influenced severely. Nearly one million Viet Namese were faced with a shortage of daily water, nearly 160,000 hm^2 of farmland was affected, and the economic loss was estimated at 10.50 million USD.[6] According to another media report,[7] the Scientific Symposium on the Environmental Impact from Hydropower Projects on the Mainstream Mekong River, held in Can Tho Province on March 4, deems 2016 witnessed the most serious alkaline erosion and drought in the Cuu Long River Plain in recent 100 years. The alkaline-eroded distance was 10 and 9 km for provinces along Tien Giang and Hau Giang. Because of climate change, aquatic product output of this region was estimated to decline by 600,000 t/a and crop output would also decline by 224,000 t/a. Total loss of agriculture and aquaculture was about 520 million VND (about 230 million USD), taking up 2.3% of its GDP.

---

[5]VN Express (March 14, 2016).

[6]UN Office for the Coordination of Humanitarian Affairs (December 15, 2016).

[7]Nhan Dan Newspaper (March 7, 2016).

Other countries attacked by severe drought included Thailand and Myanmar. In Thailand, the persistent drought for two consecutive years affected food output seriously and also influenced the economy. Nearly 70 out of the 77 provinces went through the most serious drought in decades, with more than 9.56 million people affected. More than 2 million hm^2 of farmland was damaged, and the economic loss reached hundreds of millions of USD. Myanmar suffered persistent high temperature, and agriculture and farmers there were suffering from drought.

As requested by the Vietnamese government, the Chinese government replenished water emergently to the lower reaches on March 15–May 31, 2016, during which period the Jinghong Hydropower Station increased the discharged flow from 1000 m^3/s to above 2000 m^3/s. The emergency replenishment is a spatio-temporal regulation of water resources that was realized through water storage of upper reservoir in the special conditions of lower reaches, demonstrating humanitarian spirit and positive and practical cooperation attitude of the Lancang-Mekong cooperation mechanism

2.5 Summary

Through collecting and reorganizing data in international authoritative disaster databases at and released officially by MRC and disaster bulletins these countries released publicly, we preliminarily analyzed the flood loss of the Mekong River Basin and got the following conclusions:

(1) Flood causes heavy fatalities in the basin. Flood-caused fatalities in the Mekong River Basin were most serious, 825, in 2000, followed by 2001 (489), 2011 (396), 2013 (247) and 1996 (173). The death toll is smaller than 100 in other years. Cambodia and the Cuu Long Delta, Viet Nam take the largest shares while the share is small in northeastern Thailand and Lao PDR.

(2) Flood imposes significant impact on agriculture. 2000 saw the biggest agricultural impact of floods, approximately 2.50 million hm^2, followed by 2011, about 500,000 hm^2. Seen from the geographic distribution, agricultural impact gathered in the Cuu Long River Delta, Viet Nam in 2000, and gathered mostly in Cambodia and the Cuu Long River Delta, Viet Nam and partially in Lao PDR in 2011. In 2013, floods mainly affected agriculture in the irrigation regions of northeastern Thailand.

(3) Flood incurs serious economic loss in countries at the basin. Seen from the statistical analysis of flood-caused economic loss at the basin in 1996–2014, Cambodia ranks top by 1.4 billion USD, followed by Viet Nam by 980 million USD and Lao PDR by 590 million USD. Thailand ranks bottom by 310 million USD.

(4) Extreme floods take place more frequently and cause a huge loss upon the beginning of the twenty-first century. The flood loss amounted to 1.164 billion USD in 2011, taking up 35% of the gross flood loss in 1996–2014; to 651 million

USD in 2013, taking up 20%; and to 462 million USD in 2000, taking up 14%. The flood loss was heavy in those three years. In addition, flood losses of the four countries were synchronous in time.

(5) Drought features extensive influence, long duration and huge economic loss. Though drought doesn't take place frequently, drought imposes heavy impact on population and incurs huge economic loss in these countries, and individual droughts feature extensive impact, long duration and huge economic loss and the gross loss incurred thereof tends to increase significantly. The drought loss of the Cuu Long River Delta in 2002 and May 2004 is much higher than that of the loss of 7 out of the 9 floods in 2000–2008. In northeastern Thailand, drought-caused paddy loss is about 10 million USD/year, very nearly the same with yearly flood loss in the region. Besides, drought has very apparent influence on fishery breeding at the Tonle Sap Lake, with a yearly loss of 15 million USD/year or so. Seen from spatial distribution, drought loss is heavy in northeastern Thailand, Cambodia and Viet Nam and relatively mild in Lao PDR.

Dr. Chen's research primarily focuses on comprehensive management of urban rivers, river ecological restoration, river sediment management, transboundary river water management and flood management. She was the Program Manager of many important national and provincial level projects. She has over 2 years' research experiences in transboundary water cooperation of China. The number of participated projects and programs exceeds 50. She has written 4 technology standards (or specifications) and 4 books, has published over 50 papers.

Dr. Liu's research primarily focuses on hydrological processes and modeling, flood management, and transboundary river water cooperation. She has over 5 years' research experiences in transboundary water cooperation of China. She was a participant of various Lancang-Mekong water resources cooperation activities. She is an editor of a joint publication by MWR of China and MRC in 2016. She is Program Managers of 3 international cooperation projects with Mekong River countries. She is author of over 10 journal papers and 1 book.

Dr. Baiyinbaoligao's research primarily focuses on Eco-hydraulics, urban river, and river restoration. He was the Program Manager of important national and provincial level projects. The number of investigated or participated projects and programs exceeds 70. He has written 4 technology standards (or specifications) and 4 books, has published over 80 papers.

Dr. Mu's research primarily focuses on water resources management, river ecosystem restoration and hydraulic computation. He has over 10 years' research experiences in relative national level projects. He is author of over 40 journal papers and 4 books, as well as dozens of patents for inventions.

Chapter 3
Analysis of Flood Character in the Mekong River Basin

Shiyu Hou, Hui Liu, and Fuqiang Tian

Abstract Mekong River Basin (MRB) has suffered huge injuries and property losses from frequent floods, which would very likely witness an increase of flood magnitude and frequency under climate change. In this chapter, we set up a distributed hydrological model (THREW) to provide fundamental analysis of the flood characteristics in the MRB, the simulation period is 1991–2016, the spatial coverage is the whole basin except the delta region due to lack of reliable topographic data. Two main types of flood in the MRB which are riverine flood and flash flood are discussed. Flood peak frequency at mainstream stations along Mekong are achieved by Pearson-III Frequency Curve Fitting. The annual flood volume and duration at mainstream stations along Mekong River are calculated and analyzed. Taking the flood volume of damaging floods at Stung Treng station (at lower reach of Mekong mainstream) as subject, the THREW model is used to analyze flood's travel time and regional composition, which would benefit flood prevention and water resources management from a whole-basin view.

3.1 Data and Methodology

3.1.1 Data

(1) The historical long-sequence section flow data of major hydrological stations of Chiang Saen, Luang Prabang, Nong Khai, Nakhon Phanom, Mukdahan, Pakse,

S. Hou (✉) · F. Tian
Tsinghua University, New Hydraulic Building, Beijing 100050, China
e-mail: housy16@mails.tsinghua.edu.cn

F. Tian
e-mail: tianfq@mail.tsinghua.edu.cn

H. Liu
China Institute of Water Resources and Hydropower Research, A-1 Fuxing Road, Haidian District, Beijing 100038, China
e-mail: wuyouliuhui@163.com

H. Liu (ed.), *Flood Prevention and Drought Relief in Mekong River Basin*, Springer Tracts in Civil Engineering,
https://doi.org/10.1007/978-981-15-2006-8_3

Stung Treng, Kratie on mainstream Mekong River (from upper to lower reaches, see Fig. 3.1 and Table 3.1) was collected from Mekong River Commission. The sequence length is 1985–2016, and temporal resolution is daily.

(2) As input of hydrological modelling, meteorological data, soil data and vegetation information was collected and applied in this chapter. Including:

- Ground observation meteorological/rainfall data from Mekong River Commission, including daily rainfall information from 502 rainfall stations of

Fig. 3.1 Hydrological stations along Mekong mainstream

Table 3.1 Main properties of hydrological stations along Mekong mainstream

No.	Station	Location		Country	Start/end year
		Latitude	Longitude		
1	Chiang Sean	20.274	100.089	Thailand	1985–2016
2	Luang Prabang	19.893	102.134	Lao PDR	1985–2016
3	Nong Khai	17.881	102.732	Thailand	1985–2016
4	Nakhon Phanom	17.425	104.774	Thailand	1985–2016
5	Mukdahan	16.583	104.733	Thailand	1985–2016
6	Pakse	15.100	105.813	Lao PDR	1985–2016
7	Stung Treng	13.533	105.950	Cambodia	1985–2016
8	Kratie	12.481	106.018	Cambodia	2005–2016

which 32 are conventional meteorological stations offering daily meteorological elements such as barometric pressure, temperature, humidity, wind speed and direction, sunshine duration and solar radiation etc. The period is 1991–2005.

- Satellite-based rainfall product TRMM3B42_7.0[1] was used for the period 1998–2016.
- Soil data was based on the global earth database provided by FAO (http://www.fao.org/home/en/) of which the spatial resolution is 10 km.
- As to vegetation information, NDVI and LAI data products of MODIS were used and NDVI data of Mekong Basin were downloaded from the technical website of NASA (http://reverb.echo.nasa.gov/reverb).

3.1.2 Methodology

Recognizing there are two types of flood in the Mekong River Basin, and realizing the fact of lacking first-hand local information on the flash flood, we made a brief introduction of the two types of flood, and a deep analysis of the mainstream riverine flood. For the mainstream riverine flood,

[1]Satellite TRMM, jointly developed and designed by NASA (National Aeronautics and Space Administration) and JAXA (Japan Aerospace Exploration Agency), is mainly used to monitor and study rainfall in tropical regions, actually covering a wide region ranging between 50° N and 50° S on Earth. Satellite TRMM, a low earth orbit satellite at the dip angle about 35°, launched on November 28 1997 in Japan, is the first meteorological satellite specifically used to observe rainfall in tropics and subtropics, carrying sensors such as TMI, PR, VIRS, LIS and CERES. Of devices carried by satellite TRMM, PR is ground breaking designed and made by Japan's National Space Development Agency (NASDA) and can provide 3D structure of rainstorm, capable of increasing accurate estimation of rainfall. TRMM provides rainfall data ranging between 50° N and 50° S and multiple time intervals in the globe, supplementing rainfall information of no-information regions in the globe.

(1) Flood peak, volume and duration

The characteristics of peak, volume and duration of the mainstream flood is analyzed based on historical records and observed hydrological data. Pearson-III Frequency Curve Fitting model was applied to reveal the frequency character of flood peak.

(2) Flood composition

The flood composition analysis is mainly based on THREW model (a hydrological model developed by Tsinghua University, the introduction of the model is in Appendix), and a statistic of flood travel time by corresponding discharge method.

3.2 Setting Up of the THREW Model

Please be noted that though aiming at setting up a hydrological model over the whole Mekong River basin, the delta region is still missed due to flat terrain and complex river network in the Mekong delta.

3.2.1 Forcing Data

(1) **Precipitation**

To extend the timespan coverage of the model as long as possible, two kinds of precipitation data were collected: information from meteorological/rainfall stations and satellite-based rainfall products. Ground observation data come from Mekong River Commission, including daily rainfall information from 502 rainfall stations of which 32 are conventional meteorological stations offering daily meteorological elements such as barometric pressure, temperature, humidity, wind speed and direction, sunshine duration and solar radiation etc. The distribution of the rainfall stations and the meteorological stations are shown in Fig. 3.2. The data covers the period of 1991–2005. Based on meteorological stations, Thiessen Polygons[2] was used to calculate rainfall at each representative unit basin in the distributed model. Satellite-based rainfall product TRMM3B42_7.0 was used for the period 1998–2016.

It is known from relevant researches that there are many south-north mountains in Mekong Basin on both banks as a result of that the windward slope on the left

[2]Thiessen Polygons is a method worked out by Holland climatologist Thiessen to calculate mean rainfall on the basis of rainfall of discretely distributed meteorological stations, i.e. connect all adjacent meteorological stations into triangles, draw perpendicular bisectors on each side of these triangles, and connect the crossover points (i.e. center of circumcircle) of three sides of every triangle to obtain a polygon. The rainfall intensity of the only meteorological station within the polygon is used to express the rainfall intensity within the polygon region and such polygon is called Thiessen Polygon.

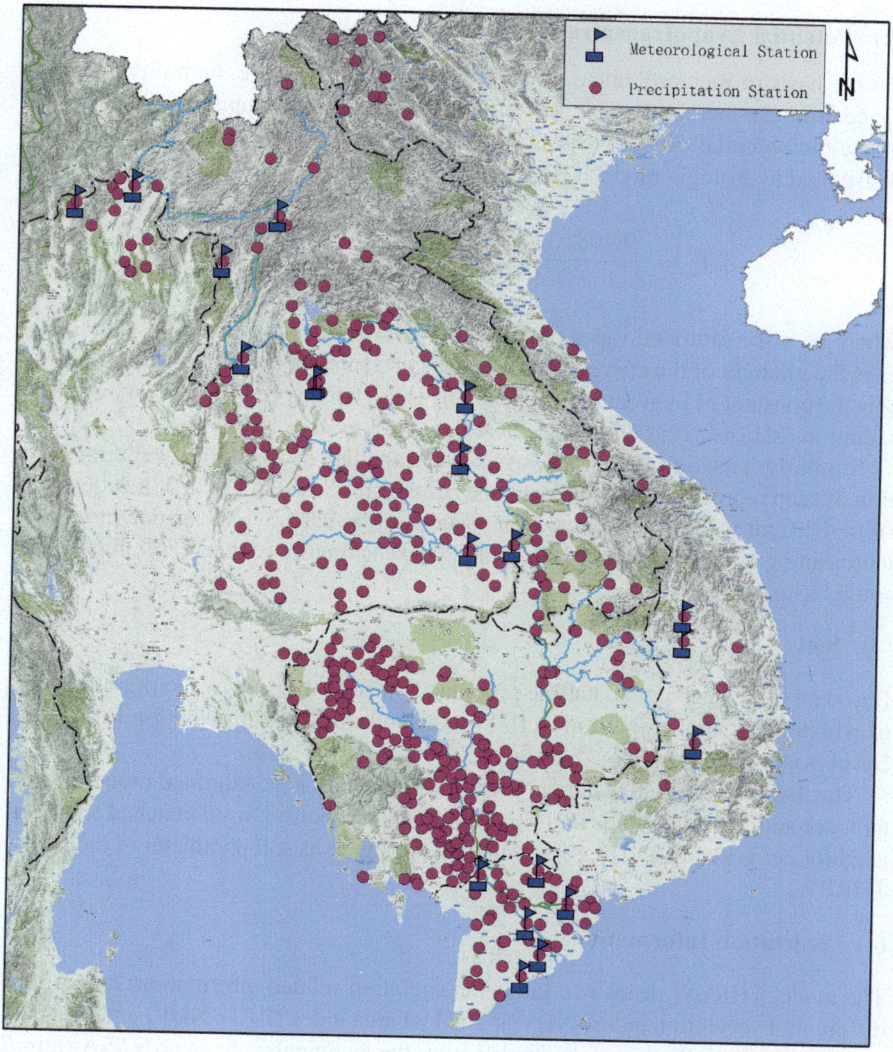

Fig. 3.2 Distribution of meteorological stations and rainfall stations in Mekong Basin

bank is a high rainfall and runoff area. While the leeward slope on the right bank (because of foehn effect) is a low-value area. In addition, within the basin, there are some mountains which are not giant also become high rainfall and runoff areas, for instance, Mount Douluo region in the lower Mekong Basin, and some giant upheaval-like landforms, for instance, Highland Nakhon are low-value areas. In the north of Chiang Saen, there are many high mountains and valleys, water vapor is hard to enter for the deep valley and larger height difference, and the foehn effect, so these areas become the low rainfall and runoff areas. The rainfall distribution obtained from the research also reflects aforementioned spatial distribution characteristics.

(2) **Potential Evapotranspiration**

On the basis of meteorological information about temperature, humidity and wind speed from 32 meteorological stations, Penman-Monteith equation is used to calculate the potential evapotranspiration and then are used as the forcing data of the distributed hydrologic model THREW. The equation is:

$$\mathrm{ET}_0 = \frac{0.408\Delta(R_n - G)}{\Delta + \gamma(1 + 0.34u_2)} + \frac{\frac{900}{T+273}\gamma u_2(e_s - e_a)}{\Delta + \gamma(1 + 0.34u_2)} \tag{3.1}$$

where, e_s is the saturated vapor pressure (kPa); e_a is the actual vapor pressure (kPa); Δ is the gradient of the curve saturated vapor pressure versus temperature (kPa/°C); γ is the constant of hygrometer (kPa/°C); u_2 is the wind speed (in m/s); R_n is the net radiation (MJ/(m^2/d)); G *is* surface heat flux (MJ/(m^2/d)).

Similarly, Thiessen Polygons was used to calculate the potential evapotranspiration of each representative unit basin, the results are shown in Fig. 3.3. It is observed from the figure that the potential evapotranspiration has obvious spatial variability, increasing gradually from north to south, and reaching more than 1200 mm/year in south basin.

(3) **Soil**

Based on the global earth database provided by FAO (http://www.fao.org/home/en/) of which the spatial resolution is 10 km, the soil thickness distribution in Mekong Basin is shown in Fig. 3.4.

The data such as field water capacity, wilting coefficient, saturated moisture content, porosity, air entry value and pore-diameter distribution, were matched and averaged for per grid unit and representative unit basin, to as soil parameter of the model THREW.

(4) **Vegetation Information**

The model THREW needs two kinds of vegetation-related information: normalized differential vegetation index (NDVI)[3] and leaf area index (LAI). NDVI and LAI data products of MODIS were downloaded from the technical website of NASA (http://reverb.echo.nasa.gov/reverb), the NDVI data of Mekong Basin in July 2000 is shown in Fig. 3.5, and it is seen from the figure that the whole Mekong Basin has higher NDVI value while Chi-Mun basins in Thailand have lower NDVI value.

[3]The value range of NDVI is $-1 \leq \mathrm{NDVI} \leq 1$, where the negative means the surface is covered with cloud, water and snow etc., high reflectance against visible light; 0 means there are rock or bare soil etc.; positive means it is covered with vegetation, increasing with increasing coverage.

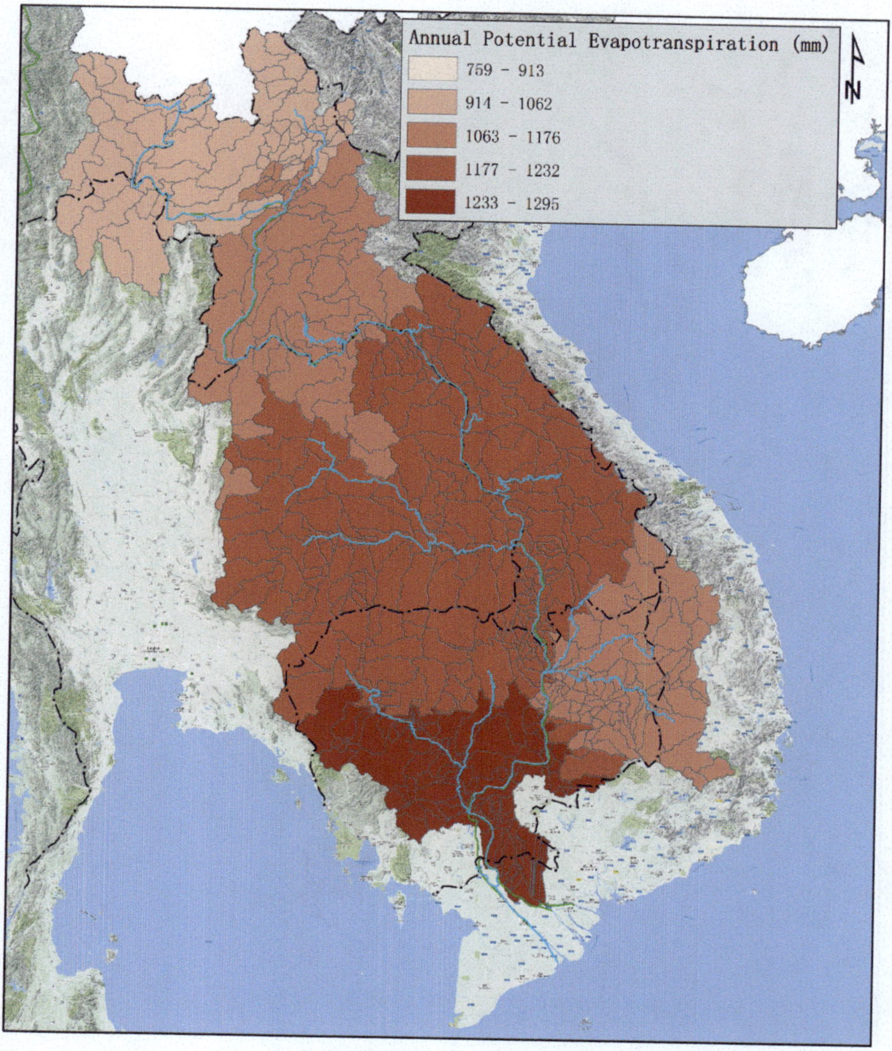

Fig. 3.3 Distribution of annual mean potential evapotranspiration in Mekong Basin (based on observation data of rainfall stations in 1991–2005)

3.2.2 Calibration and Validation

Lacking of tributary observed hydrological data, the THREW model is calibrated based on mainstream hydrological data with an upper to lower order to make sure each section of the basin is fully calibrated. During model calibration, the Nash efficiency coefficient (NSE) of different combinations of parameters were optimized from 1991 to 2000, and the combination of parameters that made the best NSEs was picked as calibration result. The validation period is 2001–2005.

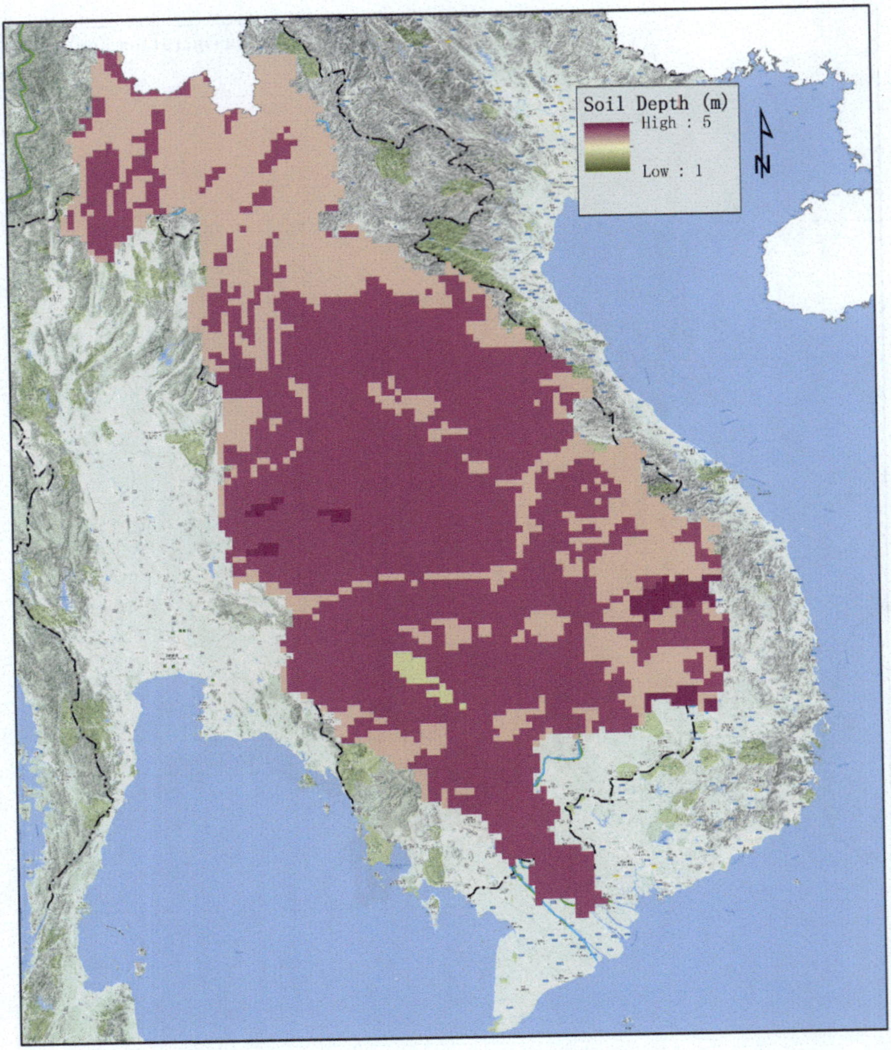

Fig. 3.4 Soil thickness distribution in Mekong Basin (plotted based on FAO's global earth database)

Table 3.2 shows NSEs calculated at each stations in 1991–2005 and it is known that the simulation results are good as a whole. 7 hydrological gauges were calibrated and verified here, and each gauge was simulated for 15 years, and a total of 105 NSE results were obtained. Among them, 62 NSEs exceeded 0.8, 82 NSE results were more than 0.7, and most low NSEs are concentrated in years of small simulated runoff. For the validation period 2001–2005, 23 out of 35 NSEs are larger than 0.7 and 30 NSEs are larger than 0.5, which means the model performance is good or acceptable; 2 out of 35 is approximating 0.5; and the left 3 is 0.2 and less, which is not acceptable, further discussion of these low NSEs is in Sect. 3.2.3.

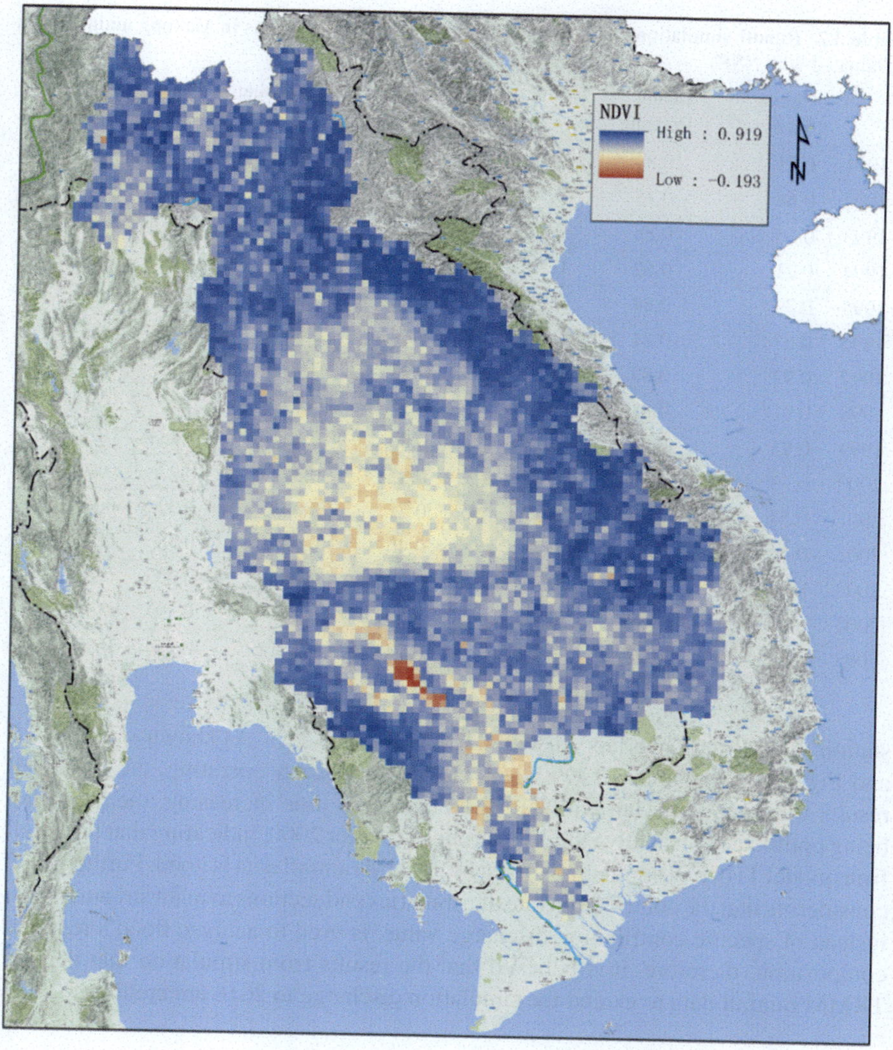

Fig. 3.5 NDVI distribution in Mekong Basin (July 2000)

Figure 3.6 is the comparison diagram between the observed runoff and simulated runoff at hydrologic stations in Mekong mainstream in 2000, where all NSEs are over 0.75 of which three stations' NSE is over 0.9, getting a good simulation result in general.

Since it is in the period 1991–2005 that observed rainfall data at stations were collected, to further extend the simulation period, TRMM rainfall data is used to extend it to 2016. To verify the quality of TRMM rainfall data, the research calculates NSE of TRMM simulated discharge using the simulated station results as reference by comparing the discharge simulation results in the period 1998–2005 when the

Table 3.2 Runoff simulation result with the model THREW on stations in Mekong main stream (evaluated with NSE)

	Chiang Saen	Luang Prabang	Nongkhai	Nakhon Phanom	Mukdahan	Pakse	Stung Treng
1991	0.81	0.84	0.84	0.83	0.90	0.87	0.87
1992	0.68	−0.15	0.50	0.63	0.83	0.82	0.87
1993	0.56	0.88	0.87	0.78	0.79	0.75	0.76
1994	0.71	0.62	0.79	0.91	0.94	0.88	0.91
1995	0.79	0.84	0.90	0.92	0.91	0.80	0.86
1996	0.67	0.84	0.86	0.89	0.86	0.73	0.83
1997	0.92	0.89	0.90	0.94	0.96	0.90	0.92
1998	0.91	0.88	0.85	0.91	0.81	0.52	0.46
1999	0.93	0.74	0.76	0.80	0.76	0.48	0.72
2000	0.78	0.93	0.91	0.82	0.89	0.92	0.89
2001	0.57	0.86	0.89	0.71	0.84	0.92	0.88
2002	0.69	0.84	0.7	0.46	0.63	0.78	0.79
2003	0.00	0.89	0.81	0.20	0.65	0.79	0.73
2004	0.69	0.89	0.88	0.52	0.71	0.79	0.81
2005	0.82	0.92	0.90	−0.04	0.43	0.56	0.87

station rainfall data overlapped with TRMM rainfall data. As shown in Table 3.3 and Fig. 3.7, in 1998 and 1999 when TRMM just began operating, the simulated results were not good, but with TRMM's operating time increasing, the algorithm being optimized constantly, NSEs are around 0.9 after 2000, indicating that the result from model THREW simulation driven by TRMM rainfall data is good. Furthermore, considering that the contribution rate of tributaries and sections to main stream floods, instead of specific contribution discharge value, is used to analyze flood's regional composition, therefore, it is believed that the results from simulation that applied TRMM rainfall data to extend the simulation discharge to 2016 are credible.

3.2.3 Discussion

(1) At the time of the calibrating, at Luang Prabang, if the NSE was high in 1991, the NSE in 1992 was low, and vice versa. Considering that the NSE should be raised as much as possible in most years, the balance was taken and parameters of the 1991 NSE high was selected finally. The reason for the low NSE in 2003 at Chiang Saen and in 2005 at Nakhon Phanom may be due to the increasing human consumption water in the relatively dry years. but there is no definite evidence to prove it.

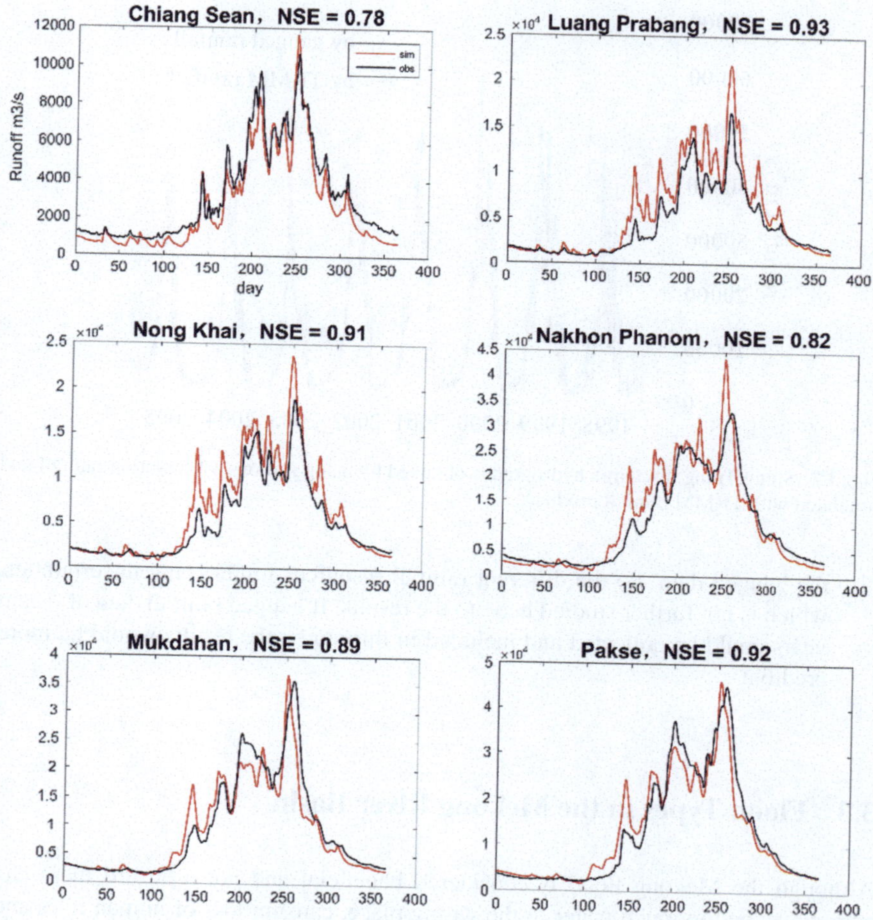

Fig. 3.6 Simulated and observed runoff NSEs in 2000

Table 3.3 NSE of Stung Treng discharge simulated with TRMM rainfall

Year	1998	1999	2000	2001	2002	2003	2004	2005
NSE	0.53	0.46	0.89	0.87	0.97	0.88	0.94	0.96

(2) Since observed gauge rainfall data was of 1991–2005, to further extend the simulation period, TRMM rainfall data was used to extend to 2016. Though the runoff simulation results indicate that simulation driven by TRMM rainfall data is credible for the flood composition analysis, considering the spatial distribution of TRMM rainfall data is different with gauge data, we found that the ratio of rainfall on the right bank to the left bank of the Mekong is bigger than that of

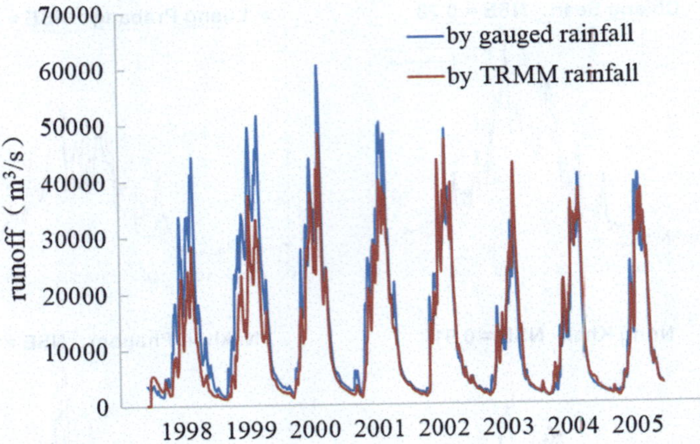

Fig. 3.7 Stung Treng discharge hydrographs obtained by using stations' observation rainfall and simulated with TRMM rainfall product

the gauged data. So the different rainfall resources might bring uncertainties, which is not further studied here, to the results. If gauged rainfall data of recent years could be collected and included in this study, the results would be more credible.

3.3 Flood Types in the Mekong River Basin

Although the Mekong flood is considered beneficial and not a hazard there can sometimes be too much water in the wrong place, causing loss of human lives and damage. But it is important to distinguish between the types of floods causing the hazards and understand the mechanism behind them, to be able to prepare for them and mitigate the impacts. An obvious example is when extreme water levels in the mainstream Mekong cause overflow of river banks and inundation of areas along the river. Such floods have been preceded by a rising water level during a period of days, and have thus given an alarm in advance. Given reliable weather and rainfall forecasts this type of flood can be predicted with a high degree of confidence to make appropriate preparations to combat it. This type of flood recedes in a few days, but can become quite prolonged in the flood plains of Cambodia and Viet Nam. On the other extreme, flash floods in the tributaries are caused by intense rainfall during a short period of time, causing high rise of water levels within a few hours. They occur in steep streams or small tributaries and the high speed of the water flow causes erosion, land and mud slides and other damage, sometimes with human casualties. They are not necessarily preceded by constant rainfall during a long time, but can occur at any time during the rainy season, and they will recede rapidly. These floods are much harder to predict, as the extreme, intense rainfall may be concentrated to a small area in the watershed. It is often the case, though, that local damage of floods often happens in connection with storms sweeping in over the Mekong Basin, bringing rainfall that in some cases last several days. Soil conditions first become saturated and then an intense but not necessarily an extreme rainfall event can rapidly trigger a flood in small watersheds, as the soil cannot absorb any more of the rainfall. The prediction of such floods calls for skills and systems especially designed for this.

Flash floods are potentially dangerous not only because water levels rise quickly, but also because water flow velocities are high. The force of the water can cause erosion, mudslides, uproot trees and tear away boulders, and the flow of debris can sweep away houses and destroy bridges. The collection of debris in the river bed should be cleared away as soon as possible to ensure that it does not block the water flow should there arrive a new flood event. Therefore it is important that disaster management agencies identify locations where flash floods and damages have occurred, to provide immediate relief and aid. As flash floods are local in character this information has in most cases to come from local authorities or village people themselves. Recovery works, such as repairing damaged roads, bridges and other infrastructure, usually needs broader support from national agencies.

As river valleys often are the most productive parts of the land it is natural that there is pressure to establish settlements in these areas, despite the risk of floods. Risk management, such as land use planning is the key to lessen the exposure to risks, and at least public structures, such as schools, hospitals and the like should be built in areas safe from floods.[4].

Since the flash floods only occur in the local watershed, the small area of the basin increases the difficulty of the simulation. In addition, the data on the flash floods at hand is limited, so it is difficult to carry out detailed analysis. Therefore, this chapter mainly analyzes the characteristics and composition of the mainstream floods.

3.4 Peak, Volume and Duration

3.4.1 Flood Peak

Based on the available discharge data of hydrological stations along Mekong mainstream, the annual maximum flood peak along the Mekong mainstream is illustrated in Fig. 3.8 and Table 3.4. It could be seen, for the upper reach (Chiang Saen, Luang Prabang and Nong Khai), the maximum flood peak occurred in 2002 and 2008; for the middle reach, the maximum flood peak happened in 1991, 1996, 1997, 2000, 2001 and 2011 (Daily discharge data of Kratie is only available for 2005–2016, which timespan is too short for frequency analysis, so flood character at Kratie is not further analyzed in this section).

[4]MRC (2015) Annual Mekong Flood Report 2014, Mekong River Commission, 70 pages, p 4.

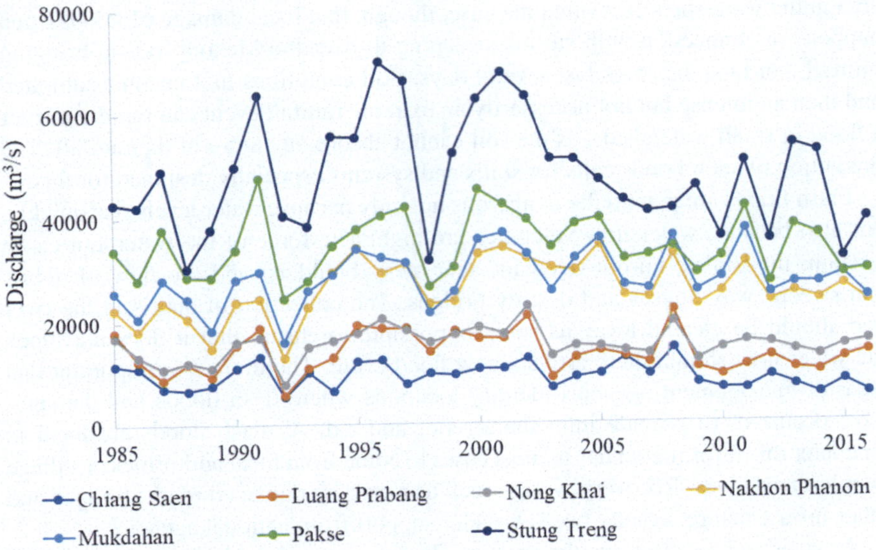

Fig. 3.8 Annual maximum flood peak along Mekong mainstream

Pearson-III Frequency Curve Fitting model is applied to reveal the frequency character of flood peaks along Mekong mainstream. The results at main hydrological stations are plotted in Figs. 3.9, 3.10, 3.11, 3.12, 3.13, 3.14, and 3.15.

3.4.2 Flood Volume and Duration

Caused by rainfall associated with the Southwest Monsoon, floods in Mekong shows a remarkable regularity, the flood season normally starting in June and ending in early November. But of course this seasonal pattern may be stronger or weaker, starting and ending earlier or later and carry different water volumes, making the individual floods having their own features. The onset of the flood season can be defined as the date when the rising discharge of the river exceeds the long-term average annual discharge. The end of the flood season is defined in a similar way as the date when the falling discharge crosses the long-term average discharge. Based on daily discharge data at mainstream stations from 1985 to 2016, the flood season duration and corresponding volume is calculated (Tables 3.5 and 3.6). Box plots of the flood volume and flood duration at mainstream stations are shown in Figs. 3.16 and 3.17. The flood volume shows an increasing trend from upstream to downstream, with 54 km^3 at Chiang Saen and 306 km^3 at Stung Treng. The flood duration varies among years and from stations, with an average ranging from 128 days to 135 days.

Table 3.4 Annual maximum flood peak along Mekong Mainstream. Unit: m^3/s

Year	Chiang Saen	Luang Prabang	Nong Khai	Nakhon Phanom	Mukdahan	Pakse	Stung Treng
1985	11,700	18,200	17,200	22,700	25,900	34,000	42,878
1986	11,000	12,600	13,300	18,400	20,800	28,300	42,196
1987	7120	8760	10,900	24,000	28,200	37,900	49,020
1988	6620	11,900	12,100	20,900	24,500	28,800	30,532
1989	7590	9450	11,100	15,100	18,500	28,700	37,950
1990	11,000	15,600	16,100	23,900	28,300	34,100	49,752
1991	13,300	18,900	17,000	24,500	29,800	47,600	63,560
1992	5370	5680	7760	13,100	16,100	24,600	40,464
1993	10,100	11,000	15,100	22,900	26,000	27,900	38,356
1994	6419	13,125	16,275	29,645	29,188	34,699	55,534
1995	12,200	19,369	17,753	34,200	34,016	37,827	55,341
1996	12,500	18,732	20,500	29,400	32,400	40,342	69,803
1997	7720	18,754	18,300	31,200	31,400	41,847	66,119
1998	9810	14,119	15,900	23,500	21,700	26,797	31,763
1999	9560	18,135	17,400	24,000	25,700	30,467	52,222
2000	10,700	16,740	18,800	33,000	35,700	45,149	62,540
2001	10,700	16,644	17,500	34,200	36,800	42,319	67,552
2002	12,700	20,943	22,800	32,100	32,900	39,343	63,049
2003	6880	8960	13,200	30,300	25,800	34,159	50,908
2004	8910	12,896	15,100	29,700	31,900	38,510	50,826
2005	9580	14,145	14,900	34,500	35,700	39,560	46,751
2006	12,724	13,050	14,051	24,857	26,837	31,760	42,923
2007	9462	11,846	13,480	24,365	26,109	32,567	40,655
2008	14,395	21,841	19,595	32,683	36,005	34,099	41,171
2009	7658	10,735	12,245	22,880	25,615	29,042	45,421
2010	6590	9326	14,471	25,425	28,082	32,111	35,846
2011	6493	11,043	15,769	31,065	37,236	41,422	50,188
2012	8941	12,842	15,309	23,600	24,939	27,175	35,287
2013	7115	10,459	13,892	25,282	26,952	37,892	53,480
2014	5412	9905	12,846	23,877	26,914	36,119	51,964
2015	7936	12,346	14,775	26,506	28,674	28,747	31,743
2016	5427	13,627	15,643	23,359	24,086	29,375	39,119

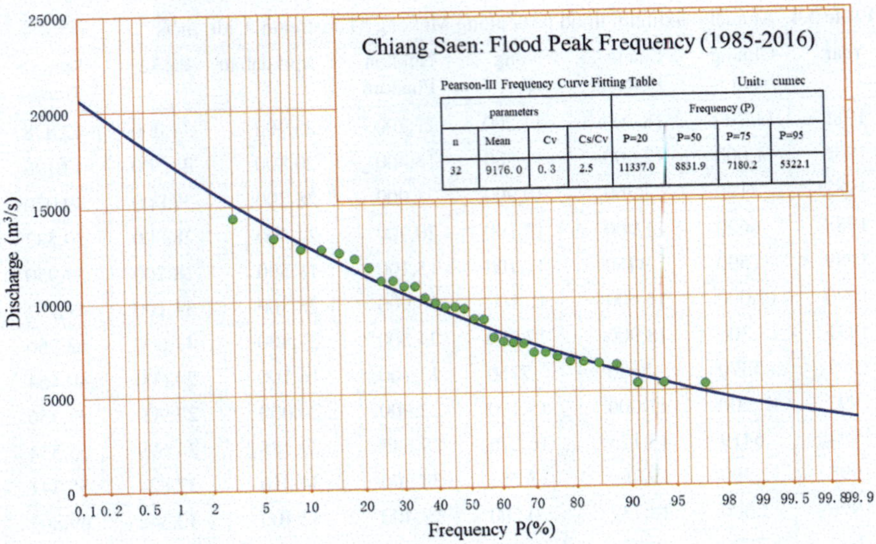

Fig. 3.9 Pearson-III Frequency Curve of flood peak at Chiang Saen

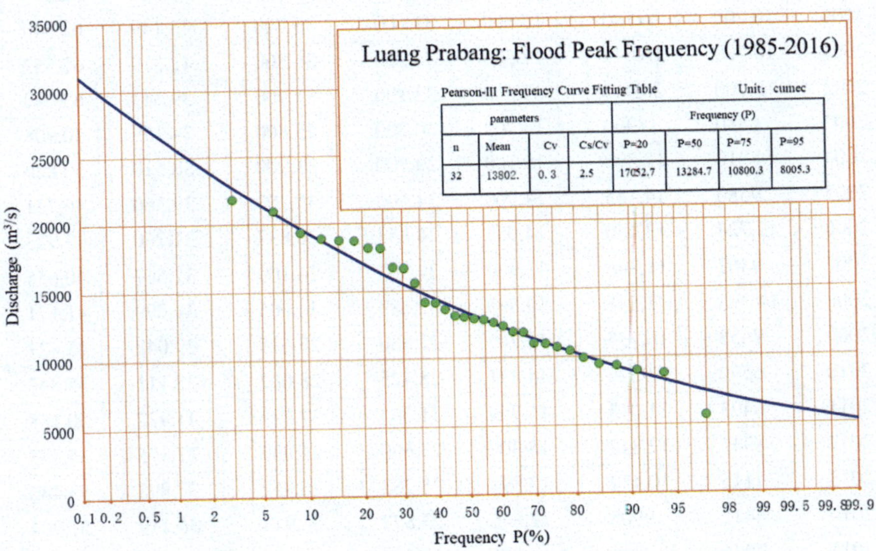

Fig. 3.10 Pearson-III Frequency Curve of flood peak at Luang Prabang

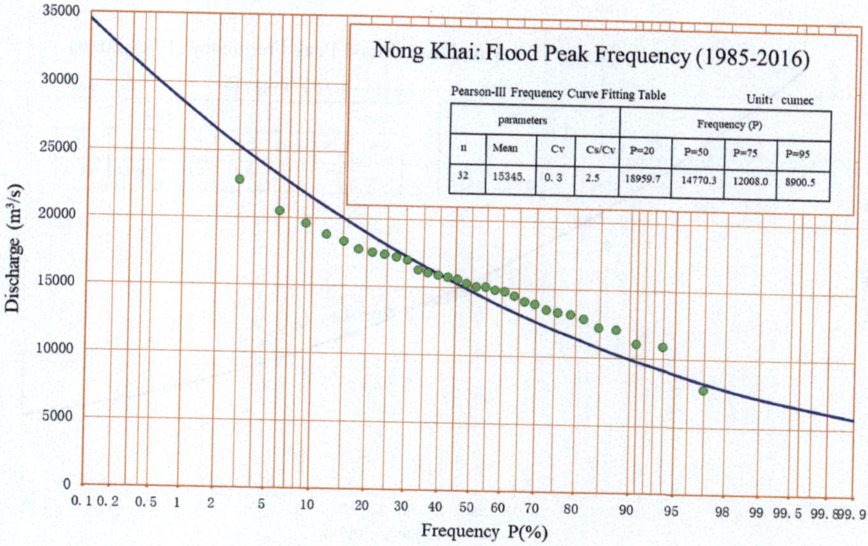

Fig. 3.11 Pearson-III Frequency Curve of flood peak at Nong Khai

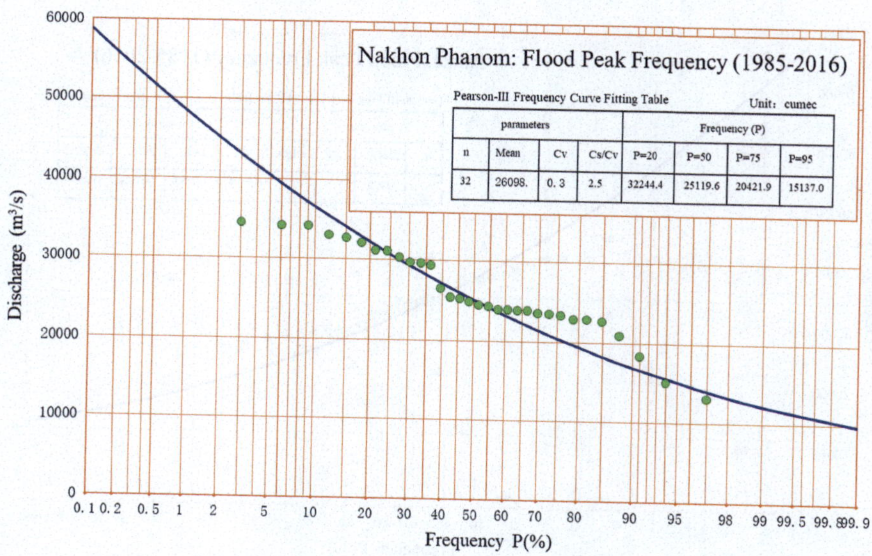

Fig. 3.12 Pearson-III Frequency Curve of flood peak at Nakhon Phanom

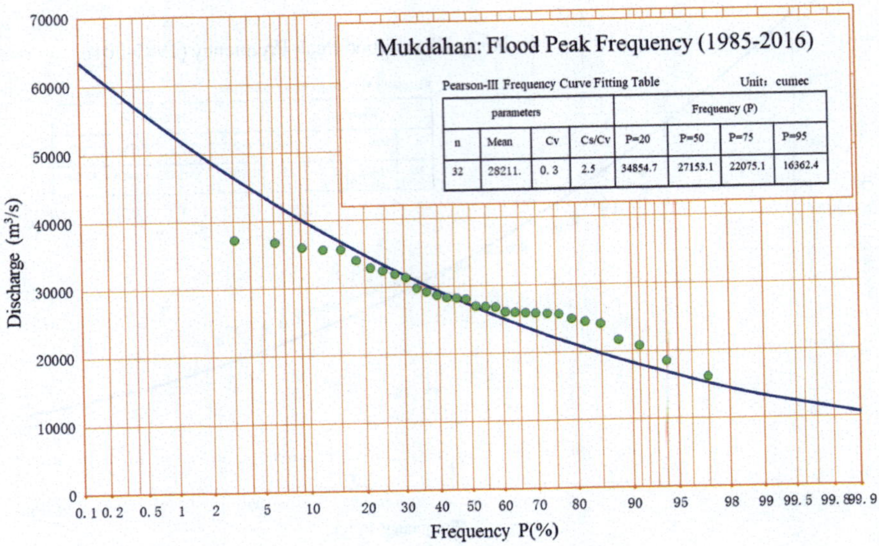

Fig. 3.13 Pearson-III Frequency Curve of flood peak at Mukdahan

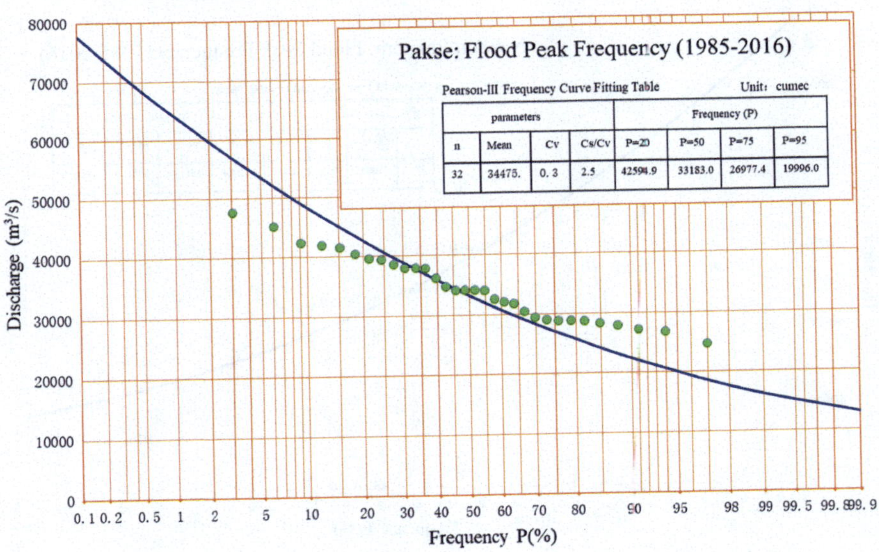

Fig. 3.14 Pearson-III Frequency Curve of flood peak at Pakse

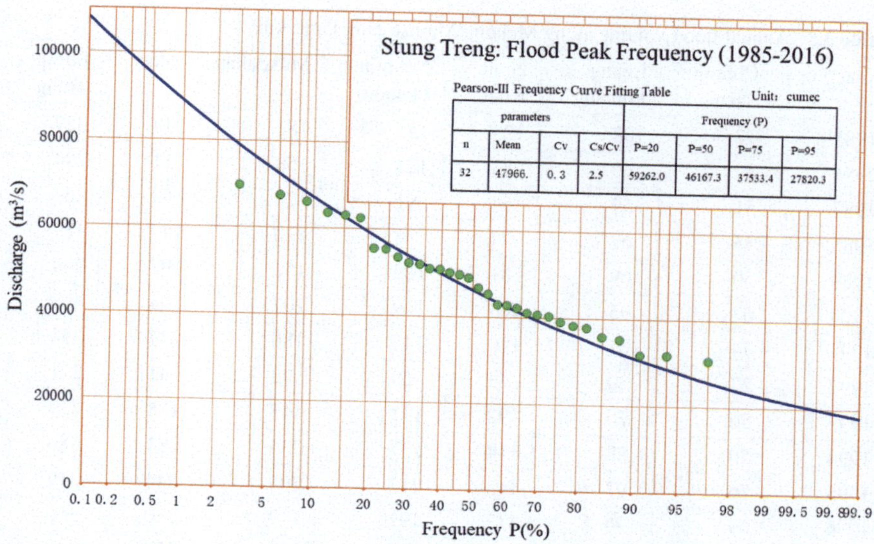

Fig. 3.15 Pearson-III Frequency Curve of flood peak at Stung Treng

3.5 Composition of Mainstream Flood

3.5.1 *Flood Travel Time*

By observing initially typical floods in recent years, i.e. the day-by-day runoff data in 2000 and 2011, it is found that in 2000 a better corresponding relation exists for day-by-day runoff of all stations while in 2011 all stations' discharge hydrographs and trends differ to some degree, suffering bigger impact of local tributary inflow, therefore calculation of flood travel time based on 2000's flood is mainly on the basis of measured data of 8 hydrologic stations along the Mekong main stream, as shown in Fig. 3.1 and Table 3.1. Using the typical floods in 2000 as subject, 8 hydrologic stations in Mekong main stream (from upstream to downstream they are Chiang Saen, Luang Prabang, Nong Khai, Nakhon Phanom, Mukdahan, Pakse, Stung Treng and Kratie) are selected to calculate flood's travel time (the water-level data are used in calculation at the most downstream station Kratie). The full-year discharge (water level) hydrographs of 8 hydrologic stations are shown in Fig. 3.18. It is understood from the figure that the shapes and trends of the discharge (water level) hydrographs of 8 stations are pretty similar, but that of 3 upstream stations is different from that of 5 downstream ones to some degree, mainly in a heavier flood process that happened in the middle and last ten days of July during which there are two peaks in 3 upstream stations and the 2nd peak is bigger while there is a single peak in 4 downstream stations and water rises fast but subsides slowly.

Based on the full-year discharge hydrographs shown in Fig. 3.18, three levels of flood processes are selected: large, medium and small, and the maximal daily

Table 3.5 Annual flood volume along Mekong Mainstream. Unit: km^3

Year	Chiang Saen	Luang Prabang	Nong Khai	Nakhon Phanom	Mukdahan	Pakse	Stung Treng
1985	77	108	123	151	181	248	327
1986	47	66	83	124	158	211	294
1987	51	59	64	91	120	169	234
1988	45	67	74	98	120	147	187
1989	50	69	81	122	152	197	254
1990	67	95	103	169	207	264	403
1991	74	107	110	142	180	257	351
1992	27	30	43	69	92	141	221
1993	56	70	92	131	145	178	239
1994	50	88	114	237	235	281	386
1995	76	97	113	213	205	244	319
1996	64	96	112	191	179	250	366
1997	49	85	95	195	195	250	340
1998	56	76	81	127	119	139	166
1999	61	99	114	206	207	266	371
2000	77	104	129	270	234	340	483
2001	84	124	145	292	277	329	455
2002	59	97	128	257	258	325	433
2003	41	52	67	153	141	174	271
2004	56	83	102	219	201	233	336
2005	47	70	82	251	230	269	321
2006	59	70	98	177	191	232	305
2007	61	73	88	174	189	228	282
2008	83	119	144	270	286	315	341
2009	44	59	75	160	168	216	284
2010	44	58	85	158	169	200	204
2011	46	81	118	252	275	337	406
2012	39	61	74	153	163	186	230
2013	44	73	92	178	190	251	304
2014	34	55	72	156	170	213	276
2015	27	46	60	136	144	160	179
2016	23	51	70	142	147	185	226
Average	54	78	95	177	185	232	306

Table 3.6 Annual flood duration along Mekong Mainstream. Unit: day

Year	Chiang Saen	Luang Prabang	Nong Khai	Nakhon Phanom	Mukdahan	Pakse	Stung Treng
1985	176	173	167	122	132	139	136
1986	122	119	127	117	135	140	136
1987	134	113	93	81	98	111	112
1988	125	122	116	89	98	102	106
1989	137	133	127	120	130	127	113
1990	170	165	157	146	152	155	165
1991	155	154	147	113	129	138	131
1992	98	81	93	81	92	97	110
1993	135	130	128	106	107	107	108
1994	138	139	140	143	142	140	138
1995	173	128	122	127	125	127	127
1996	141	136	121	125	104	129	132
1997	124	123	118	120	118	118	118
1998	118	105	99	97	94	95	97
1999	147	145	153	171	164	173	174
2000	162	150	159	173	148	168	170
2001	182	182	182	177	170	167	166
2002	132	127	136	148	141	152	149
2003	117	106	108	114	113	107	119
2004	132	127	139	149	130	117	130
2005	134	124	112	153	128	123	122
2006	154	118	133	124	125	127	127
2007	134	118	116	133	134	133	132
2008	174	168	168	171	170	174	165
2009	116	109	111	123	123	133	126
2010	126	110	113	110	111	114	104
2011	128	139	138	138	137	152	152
2012	106	106	103	129	128	126	119
2013	142	144	139	138	138	149	132
2014	114	108	105	118	122	122	119
2015	80	86	91	100	100	100	98
2016	78	134	136	142	143	159	157
Averagee	135	129	128	128	128	132	131

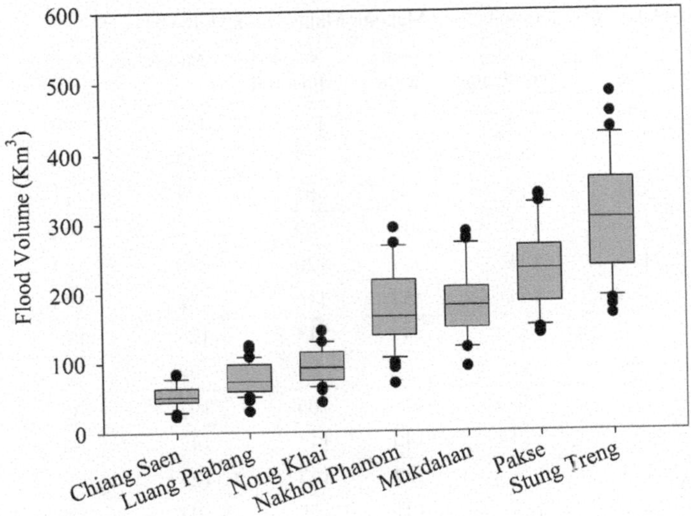

Fig. 3.16 Box plot of annual flood volume at Mekong mainstream stations (1985–2016)

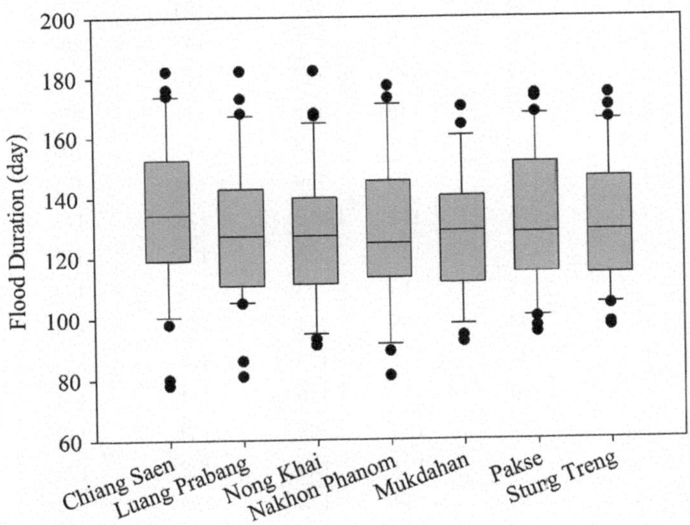

Fig. 3.17 Box plot of annual flood duration at Mekong mainstream stations (1985–2016)

discharge of each process is used as the peak discharge of the flood process to determine corresponding discharge and estimate the flood's travel time. Taking Stung Treng station as example, three flood processes occurred in the last ten days of May, the middle/last ten days of June and the first ten days of September, and the flood process that occurred in the last ten days of July is not selected as corresponding

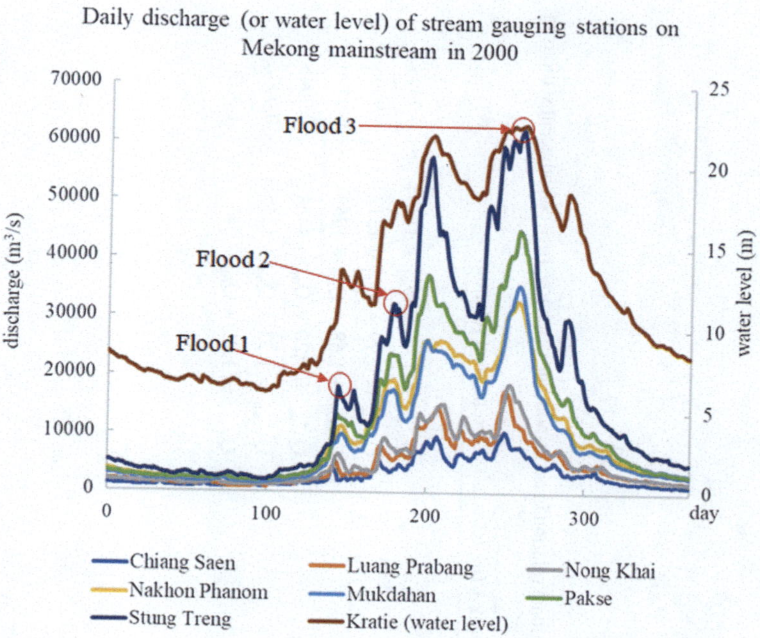

Fig. 3.18 Discharge (water level) hydrographs of hydrologic stations in Mekong main stream in 2000

discharge because there are obvious difference in discharge between different stations and its discharge level is pretty close to that of flood process #3. Table 3.7 displays the peak discharges of three flood processes of all hydrologic stations and the correlational relation in corresponding discharge between adjacent stations, and their correlational relations are shown in Fig. 3.19.

It is understood from Table 3.7 that there are good corresponding relation in 3 flood-peak discharges between stations which can be used as corresponding discharge to estimate the flood's travel time. Furthermore, there is also good correlational relation between the highest water level at Kratie and the flood-peak discharge at Stung Treng ($Z_{down} = 0.0002 * Q_{up} + 10.64$, $R^2 = 0.9697$) and therefore the water level is considered to be corresponding to the flood-peak discharges of upstream stations and can be used to estimate the travel time of the section Stung Treng–Kratie.

On the basis of the flood processes and the flood-peak discharged determined with aforementioned methods, the flood's travel time between two stations can be determined by comparing the time difference of corresponding discharges between two stations. Table 3.8 and Fig. 3.20 show the time spent by flood peaks of each discharge level traveling to hydrologic stations along the river. Because the obtained discharge data sequences are on daily scale, the temporal resolution in analysis of flood's travel time can be day only.

It is known from Table 3.8 that floods at three levels of flood-peak discharges take 5, 10 and 11 days respectively to travel from Chiang Saen station to Kratie station.

Table 3.7 Peak discharge of hydrologic stations in Mekong main stream and the correlational relation in corresponding discharge between adjacent stations in 2000

Station	Flood peak 1 (m^3/s)	Flood peak 2 (m^3/s)	Flood peak 3 (m^3/s)	Correlational relation	R^2
Chiang Saen	4350	5860	10,700		
Luang Prabang	5443	8090	17,714	$Q_{down} = 1.95 * Q_{up} - 3149$	0.9995
Nongkhai	6620	9400	18,800	$Q_{down} = 0.99 * Q_{up} + 1313$	0.9998
Nakhon Phanom	11,500	19,400	33,000	$Q_{down} = 1.68 * Q_{up} + 1753$	0.9770
Mukdahan	10,100	17,800	35,700	$Q_{down} = 1.20 * Q_{up} - 4454$	0.9945
Pakse	13,415	23,550	45,149	$Q_{down} = 1.23 * Q_{up} + 1215$	0.9996
Stung Treng	18,201	32,356	62,540	$Q_{down} = 1.40 * Q_{up} - 546$	1.0000

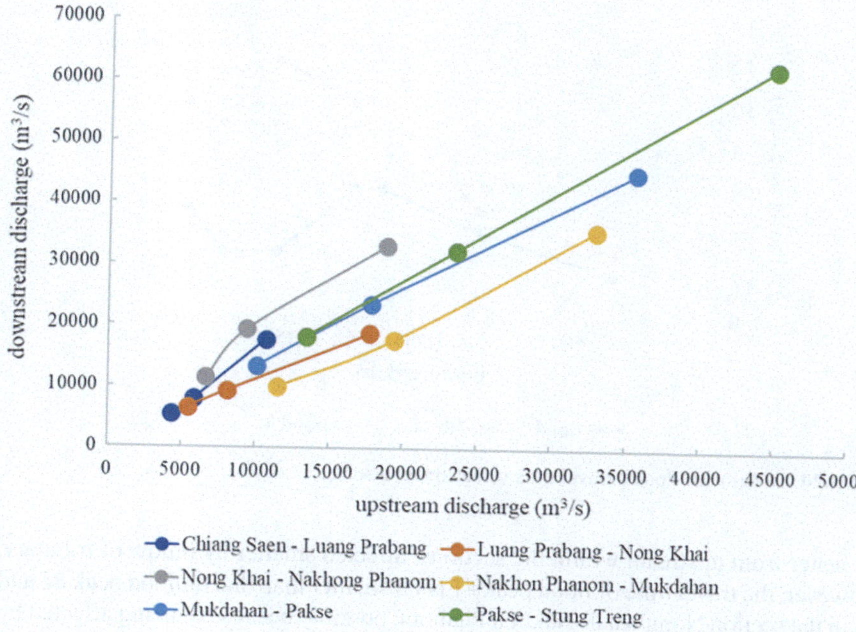

Fig. 3.19 Peak discharge of hydrologic stations in Mekong main stream and the correlational relation in corresponding discharge between adjacent stations in 2000

Table 3.8 Flood's travel time in Mekong main stream in 2000 (start point: Chiang Saen station)

Station	River length to Chiang Saen station (km)	Flood peak 1 Travel time (day)	Flood peak 2 Travel time (day)	Flood peak 3 Travel time (day)
Luang Prabang	354	2	3	0
Nongkhai	815	4	5	3
Nakhon Phanom	1143	6	10	8
Mukdahan	1236	6	10	9
Pakse	1498	3[a]	12	8[a]
Stung Treng	1681	3[a]	9[a]	10
Kratie	1804	5	10	11

[a]The flood peak at the downstream hydrologic station occurs earlier than its upstream counterpart

But the calculated results indicate that peaks at some downstream station occurred earlier than their upstream counterpart, possibly because inflow of local tributaries, which will lead to error in calculation of flood's travel time.

It is known from Fig. 3.20 that in the first 4 sections (from Chiang Saen to Mukdahan) there are good correlational relations between the flood's travel time and travel distance, so it is believed that the changes of downstream discharge are mainly caused

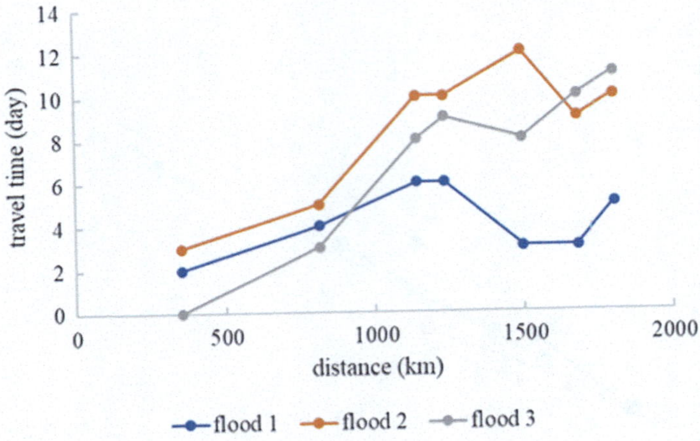

Fig. 3.20 Relation of flood's travel time versus travel distance

by water from upstream within the sections, affected smaller by inflow of tributary. However, the travel time of flood peak #1 is far smaller than that of flood peak #2 and #3 in the section Nong Khai–Nakhon Phanom, possibly because of being affected by local tributary, resulting in error in the relation of travel time versus travel distance. Therefore flood peak #1 is used to verify corresponding discharges for flood peak #2 and #3 (Table 3.7 and Fig. 3.19) only and not to further estimate the travel time. On the basis of the correlational relations between the travel time and the travel distance of flood peak #2 and #3 at river sections above Mukdahan, the time spent by flood to travel to the most downstream station Kratie and the travel speed of the flood wave are estimated, with their results shown in Table 3.9.

It is known from Table 3.9 that by using the upstream sections with good correlational relation of travel time versus distance in estimation, the time spent by flood peak #2 and #3 to travel from Chiang Saen to Kratie is 14.9 and 14.5 days respectively and the travel speed of flood wave is 1.40 m/s and 1.44 m/s respectively. According to the report Joint Assessment of Results of China's Emergency Water Make-up to Mekong released by Mekong Commission, during the emergency water make-up of China to Mekong, it takes 17 days for the making-up water (flow rate of net make-up water is about 1000 m^3/s) to travel from Chiang Saen to Kratie, close to the result

Table 3.9 Correlational relation of flood's travel time versus travel distance

Flood process	Time–distance relation	Number of days spent to arrive at Kratie	Ravel speed of flood wave (m/s)
Flood peak 2	$T = 0.0086L - 0.6136$ $(R^2 = 0.9881)$	14.9	1.40
Flood peak 3	$T = 0.0104L - 4.2551$ $(R^2 = 0.9609)$	14.5	1.44

in this report. According to the result, the bigger the flow rate is, the faster the flood travels, which is consistent with the regularity of flood travel.

3.5.2 Flood Composition

Distributed hydrological model (THREW) was applied to simulate the discharge processes of various cross-sections and tributaries, and based on continuous simulation of long sequence, the annual maxima method and the super-quantitative method are combined to select a series floods to analyze regularities on flood's regional composition.

While selecting floods, we mainly focus on floods happened at the flatter downstream section (modelled hydrograph at Stung Treng was used), and those that would possibly cause huge damages. It is also needed to determine whether flood happens and the start/end time of floods. Due to limited data, based on descriptions on flood disaster, with water level information considered, corresponding flood time periods and discharges are found out to select the heaviest floods in the period 1991–2016.

The warning water level at Stung Treng is 10.7m[5] (red line in Fig. 3.21), and the corresponding discharge is determined as 40,000 m^3/s (both flood peaks correspond to different discharges, use the lower one 40,000 m^3/s), and based on this, floods are selected: if the flood peak is more than 40,000 m^3/s, it is considered a flood

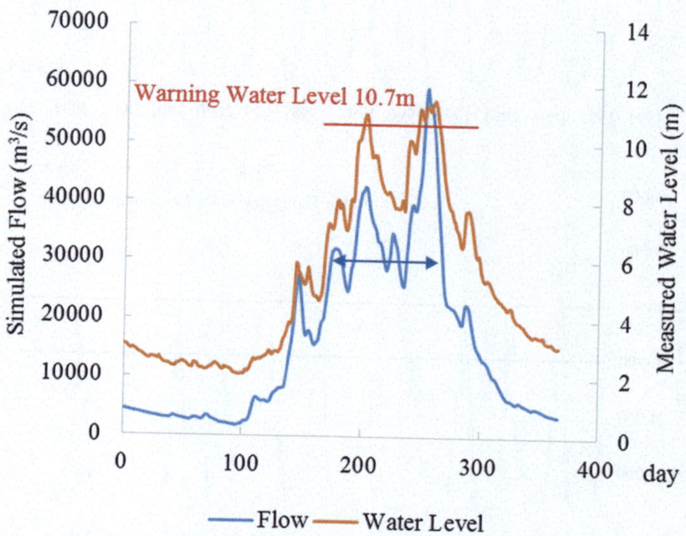

Fig. 3.21 Water level and discharge hydrograph at station Stung Treng in 2000

[5]Obtained from MRC report Seasonal Flood Situation Report for the Lower Mekong River Basin—Covering period from 1st June to 13rd November 2011.

happened. Mekong Commission's report 20 Years of Cooperation stated "in 2000, the largest flood happened in Mekong downstream 20 km distance from Phnom Penh in a few decades, which inundated villages around it and lasted 3 months." The range indicated with blue arrow in Fig. 3.21 is 98 days, consistent with 3 months described, and therefore 30,000 m^3/s is selected as threshold of the flood.

Based on the discharge hydrographs at Stung Treng station in 1991–2016 and aforementioned standards, a flood is recognized every year and its start/end date is determined, of which the flood peaks in 1992, 1993, 2003, 2007, 2009, 2010, 2012 and 2015 were smaller than 40,000 m^3/s but their flood start/end dates are determined by using the same standard as the heaviest flood of the year, thus there is a flood every year from 1991 to 2016, 26 floods in total (Fig. 3.22; Table 3.10).

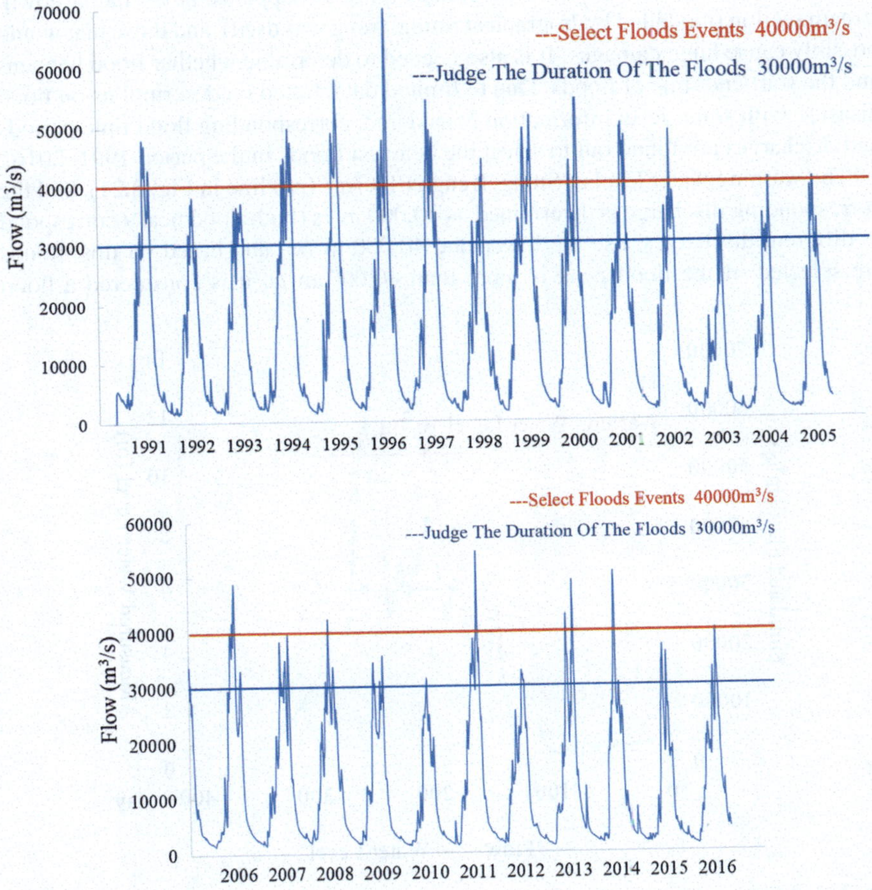

Fig. 3.22 Discharge hydrographs at Stung Treng station in 1991–2016 (results from simulation with the model THREW)

Table 3.10 Typical flood information in 1991–2016

No.	Year	Start date	End date	Duration (day)	Total flood volume (km^3)	Average discharge (m^3/s)	Remarks
1	1991	July 25	October 16	84	266	36,680	
2	1992	August 8	September 9	33	95	33,342	The flood peak smaller than 40,000 m^3/s
3	1993	July 5	September 27	85	234	31,914	The flood peak smaller than 40,000 m^3/s
4	1994	July 9	October 9	93	312	38,873	
5	1995	July 19	October 2	76	278	42,328	
6	1996	July 24	October 19	88	329	43,228	
7	1997	July 16	October 11	88	306	40,236	
8	1998	August 13	September 13	32	89	32,192	
9	1999	July 26	October 11	78	274	40,653	
10	2000	July 2	September 25	86	282	37,896	
11	2001	August 4	October 1	59	209	40,984	
12	2002	July 9	September 29	83	243	33,831	
13	2003	August 26	September 22	28	71	29,287	The flood peak smaller than 40,000 m^3/s
14	2004	August 8	September 26	50	138	31,881	
15	2005	July 28	September 24	59	168	32,888	
16	2006	August 10	September 27	49	139	32,882	
17	2007	July 19	October 19	93	269	33,445	The flood peak smaller than 40,000 m^3/s
18	2008	August 7	October 20	75	199	30,775	

(continued)

Table 3.10 (continued)

No.	Year	Start date	End date	Duration (day)	Total flood volume (km^3)	Average discharge (m^3/s)	Remarks
19	2009	July 26	September 24	61	161	30,507	The flood peak smaller than 40,000 m^3/s
20	2010	August 31	September 7	8	21	30,461	The flood peak smaller than 40,000 m^3/s
21	2011	August 2	October 11	71	226	36,884	
22	2012	August 26	September 18	24	65	31,424	The flood peak smaller than 40,000 m^3/s
23	2013	July 28	October 7	72	206	33,089	
24	2014	July 24	August 17	25	87	40,307	
25	2015	July 29	September 10	44	107	28,124	The flood peak smaller than 40,000 m^3/s
26	2016	August 21	October 3	44	116	30,459	

Table 3.11 Flood evolution time of main-stream stations and tributaries (the hydrologic stations start at Jinghong, the tributaries start at Ziqu)

Station	Corresponding time (day)	Tributary	Corresponding time (day)	Tributary	Corresponding time (day)
Jinghong	0	Ziqu River	0	Se Bang Hieng River	17
Chiang Saen	2	Angqu River	0	Mun River	17
Luang Prabang	4	Yangbi River	4	Se Done River	17
Nongkhai	8	Nanlei River	7	Sekong River	19
Nakhon Phanom	11	Nam Ou River	11	Se San River	19
Pakse	14	Nam Ngum River	13	Sre Pok River	19
Stung Treng	15	Nam Theun River	14		
Kratie	16	Songkhram River	15		

The flood travel speed fitting formula provided in the Sect. 3.5.1 is used to reckon the flood evolution time (see Table 3.11), and the model simulation discharge process is combined to calculate the contribution rate of tributary basin and local basins of hydrologic stations to Stung Treng flood volume. Figures 3.23, 3.24, 3.25 and 3.26 respectively display the contribution rates of tributaries and station-related sections to Stung Treng flood volume in 26 floods, and detailed information are present in Tables 3.12 and 3.13. It is evident that, of tributaries, Mun River, Se Done River and Tonle Sap River's flood contribution rates are maximal, and of sectional basins, Nong Khai–Nakhon Phanom, Pakse-Stung Treng, Nakhon Phanom-Pakse and Chiang Saen-Luang Prabang's contribution rates are maximal. Considering Fig. 3.27, it is known that the high runoff-generation contribution rate of Mun River and Tonle Sap River is due to large basin area and the high flood contribution rate of a few tributaries like Segong River is due to large rainfall intensity within their basins.

For all rivers, the discharge reaches maximum in summer in general except Sre Pok River and Tonle Sap River in the downstream of Mekong of which the discharge in autumn is more than in summer because monsoon retreats from north to south gradually. For tributaries near the upstream, water volume mostly reaches the maximum in August, for those near the downstream, water volume mostly reaches the maximum in September, while water volume of Tonle Sap River reaches maximum in October, so there are 3 months' peak staggering from the upstream to the downstream.

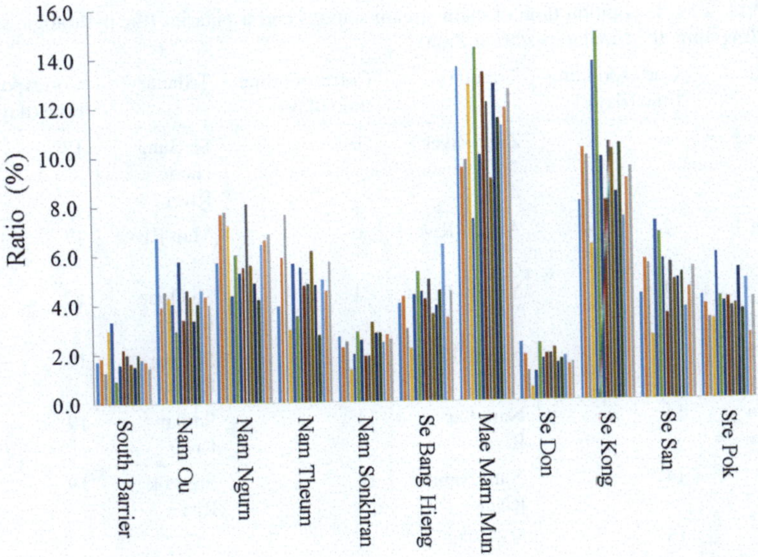

Fig. 3.23 Contribution rate of main tributaries in Mekong to Stung Treng flood volume (the results from simulation with the model THREW driven by the observed rainfall data at rainfall station from 1991 to 2005)

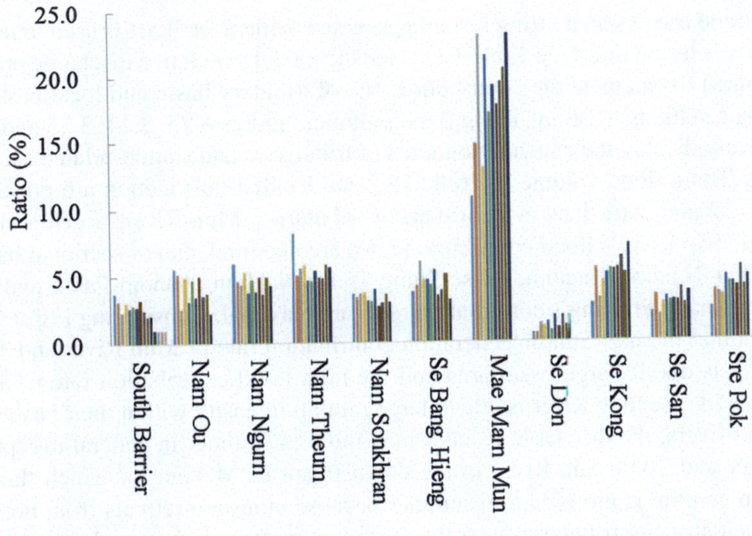

Fig. 3.24 Contribution rate of main tributaries in Mekong to Stung Treng flood volume (the results from simulation with the model THREW driven by TRMM rainfall products)

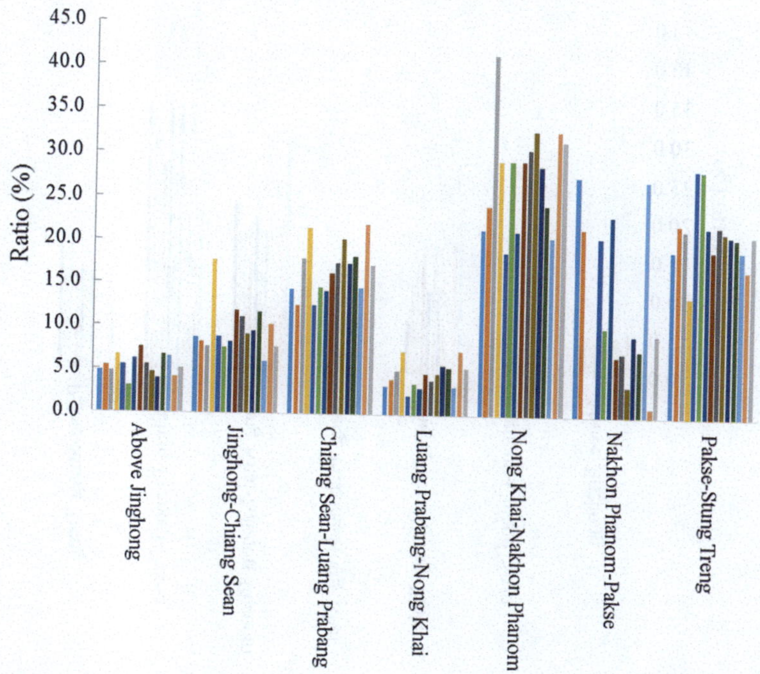

Fig. 3.25 Contribution rate of sectional basins of Mekong to Stung Treng flood volume (the results from simulation with the model THREW driven by observed rainfall data at rainfall station)

3.6 Conclusion

(1) There are two main types of flood in the Mekong River Basin. For the riverine flood, it could be predicted and effectively prevented from thereby damages with proper measures. For the flash flood, it is usually local and hard to predict. It is important for the local agencies to develop monitoring and early warning system, as well as carry out immediate relief and aid.

(2) Based on yearly flood peak analysis of 1985–2016 using observed discharge data of hydrological stations, it shows that the maximum flood peak happened in 2002 and 2008 at the upper reach of Mekong River Basin (Chiang Saen, Luang Prabang and Nong Khai), with an extreme peak of around 20,000 cumec; while that of the middle reach occurred in 1991, 1996, 1997, 2000, 2001 and 2011, with an extreme peak of 70,000 cumec at Stung Treng. Flood peak frequency at mainstream stations along Mekonog are also achieved by Pearson-III Frequency Curve Fitting.

(3) By defining the onset and withdrawal date of flood season, the annual flood volume and duration at mainstream stations along Mekong River was calculated

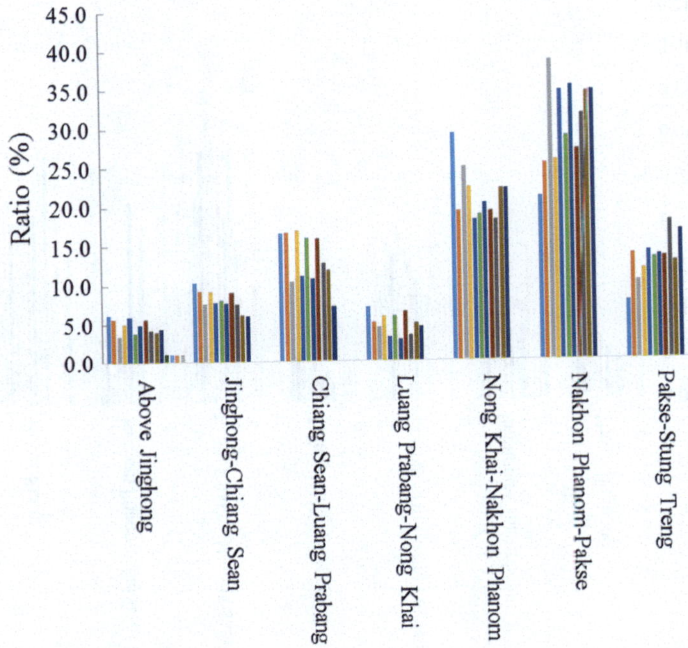

Fig. 3.26 Contribution rate of setional basins of Mekong to Stung Treng flood volume (the results from simulation with the model THREW driven by TRMM rainfall products)

and analyzed. The flood volume shows an increasing trend from upper to lower reaches, with 54 km^3 at Chiang Saen and 306 km^3 at Stung Treng. The flood duration varies among years and stations, with average ranging from 128 days to 135 days.

(4) On the basis of daily runoff data of hydrologic stations in Mekong mainstream in 2000, it takes 14.9 and 14.5 days for flood peaks at two discharge levels of 5860 m^3/s and 10,700 m^3/s to travel from Chiang Saen station to Kratie station, close to the travel time for China's emergency water supplement to Mekong in 2016.

(5) Taking the flood volume of damaging floods at Stung Treng station as subject, the model THREW is used to analyze flood's regional composition in main stream and the results indicate: of main tributaries, Mun River, Sekong River and Nam Ngum's contribution rates to Stung Treng total flood volume are highest, 15.9%, 8.6% and 6.0% respectively; of sectional basins, Nong Khai-Nakhon Phanom, Pakse-Stung Treng, Nakhon Phanom-Pakse's contribution rate are highest, 25.0%, 21.3% and 19.6% respectively.

Table 3.12 Contribution rate of main tributaries in Mekong to Stung Treng flood volume (the results from simulation with the model THREW from 1991 to 2016)

Flood	North of Jinghong	Nanlei River	Nam Ou	Nam Ngum	Nam Theun	Songkhram River	Se Bang Hieng	Mun River	Se Done	Sekong	Se San	Srepok
1	5.4	2.5	3.9	6.2	5.3	2.2	4.8	15.1	2.2	9.2	4.0	4.6
2	6.2	1.0	3.5	6.9	4.0	3.3	6.1	16.6	2.6	17.2	7.8	4.7
3	5.2	1.4	4.9	8.3	8.2	2.7	3.2	10.6	1.3	10.7	5.9	3.6
4	7.2	1.7	4.7	6.0	6.7	3.4	3.8	9.7	2.3	11.0	5.2	4.1
5	6.1	1.9	4.9	7.3	5.0	3.0	3.7	13.1	1.5	9.9	5.0	2.9
6	3.4	1.6	4.3	7.5	6.2	2.8	4.9	13.7	1.7	10.3	5.9	4.5
7	6.9	1.7	3.8	5.4	5.3	3.1	4.3	14.2	1.7	9.3	5.4	5.9
8	8.7	3.4	4.9	8.3	3.4	1.5	2.5	14.8	0.7	7.2	3.0	3.8
9	5.9	2.1	4.3	4.5	2.9	3.0	4.8	12.3	1.8	11.1	5.5	3.9
10	5.1	2.2	5.0	8.8	5.2	2.1	5.3	13.1	2.1	11.3	6.0	4.0
11	4.1	1.9	4.9	6.9	5.3	2.5	6.8	11.8	1.9	7.8	3.9	5.1
12	6.9	1.7	6.1	5.6	5.8	2.6	4.7	10.5	1.8	10.4	6.0	4.1
13	6.8	1.8	7.0	6.0	4.1	2.8	4.2	14.3	2.5	8.5	4.5	4.4
14	4.4	2.0	4.2	8.3	6.4	2.4	4.6	10.2	2.0	11.0	6.1	4.1
15	5.4	3.5	4.4	4.7	6.0	2.1	4.6	7.9	1.3	14.6	7.7	6.3
16	5.7	3.1	3.9	4.3	6.1	4.5	6.1	21.8	2.2	6.0	3.4	4.9
17	5.1	3.0	4.2	4.5	5.8	3.5	4.6	23.6	2.3	7.6	4.7	6.7
18	7.0	3.2	3.6	4.5	5.5	3.6	5.4	25.2	1.8	5.9	3.9	5.6
19	5.6	3.4	5.4	5.8	6.4	4.1	5.2	14.3	1.2	5.0	3.1	3.6
20	3.6	2.4	2.6	4.5	5.9	3.8	6.3	24.4	1.4	3.4	2.0	3.7

(continued)

Table 3.12 (continued)

Flood	North of Jinghong	Nanlei River	Nam Ou	Nam Ngum	Nam Theun	Songkhram River	Se Bang Hieng	Mun River	Se Done	Sekong	Se San	Srepok
21	4.5	2.3	4.1	5.5	6.5	4.0	5.7	22.9	1.3	5.6	3.0	4.5
22	4.3	3.3	5.3	5.3	5.2	3.4	4.8	19.4	0.9	6.0	3.3	4.8
23	6.2	3.5	5.5	5.4	5.2	3.0	3.8	19.5	1.1	5.9	3.2	4.3
24	5.1	2.4	2.6	5.0	6.5	3.5	3.6	27.0	2.2	8.5	4.3	5.6
25	6.6	4.1	6.0	6.4	8.8	3.9	3.9	11.5	0.6	3.0	2.2	2.6
26	6.6	3.6	5.9	5.2	5.8	3.9	4.8	16.8	0.7	6.1	3.9	4.2
Mean	5.7	2.5	4.6	6.0	5.7	3.1	4.7	15.9	1.7	8.6	4.6	4.5
Min	3.4	1.0	2.6	4.3	2.9	1.5	2.5	7.9	0.6	3.0	2.0	2.6
Max	8.7	4.1	7.0	8.8	8.8	4.5	6.8	27.0	2.6	17.2	7.8	6.7

Table 3.13 Contribution rate of sectional basins of Mekong to Stung Treng flood volume (the results from simulation with the model THREW from 1991 to 2016)

Flood	North of Jinghong	Jinghong–Chiang Saen	Chiang Saen–Luang Prabang	Luang Prabang–Nongkhai	Nongkhai–Nakhon Phanom	Nakhon Phanom–Pakse	Pakse–Stung Treng
1	4.8	8.7	14.4	3.3	21.5	27.7	19.7
2	5.4	8.2	12.5	4.1	24.2	21.7	24
3	4.8	7.7	17.9	5.0	41.4	−2.6[a]	25.9
4	6.7	17.7	21.5	7.3	29.4	−1.0[a]	18.3
5	5.5	8.8	12.6	2.2	18.8	20.7	31.5
6	3.1	7.5	14.6	3.6	29.4	10.2	31.6
7	6.2	8.3	14.2	3.0	21.3	23.2	23.9
8	7.6	11.9	16.2	4.8	29.4	6.9	23.3
9	5.5	11.0	17.4	4.0	30.6	7.2	24.2
10	4.7	9.1	20.2	4.8	32.8	3.4	25
11	3.9	9.5	17.4	5.7	28.8	9.3	25.3
12	6.6	11.7	18.2	5.4	24.4	7.5	26.2
13	6.5	6.0	14.6	3.3	20.6	27.2	21.8
14	4.1	10.3	22.0	7.4	32.8	1.2	22.2
15	5.1	7.6	17.3	5.4	31.6	9.4	23.7
16	5.0	7.7	10.7	2.8	20.3	35.2	18.4
17	4.2	7.5	12.6	3.4	18.0	31.5	22.8
18	6.0	7.8	11.0	3.2	18.1	34.6	19.3
19	5.1	9.3	17.0	5.9	22.3	25.7	14.7

(continued)

Table 3.13 (continued)

Flood	North of Jinghong	Jinghong–Chiang Saen	Chiang Saen–Luang Prabang	Luang Prabang–Nongkhai	Nongkhai–Nakhon Phanom	Nakhon Phanom–Pakse	Pakse–Stung Treng
20	3.4	7.6	10.4	4.4	24.8	38.4	10.9
21	4.0	6.1	11.8	4.9	22.1	34.3	16.8
22	3.8	8.2	15.9	5.9	18.7	28.8	18.6
23	5.7	9.1	15.8	6.4	19.2	27.0	16.8
24	4.4	6.1	7.2	4.5	22.0	34.5	21.3
25	6.2	10.3	16.5	7.1	29.2	21.0	9.8
26	5.8	9.3	16.7	5.0	19.3	25.3	18.6
Mean	5.2	9.0	15.3	4.7	25.0	19.6	21.3
Min	3.1	6.0	7.2	2.2	18.0	−2.6[a]	9.8
Max	7.6	17.7	22.0	7.4	41.4	38.4	31.6

[a]The contribution rate of section is negative. The fitting formula in 3.3 is used to calculate the contribution rate, and the flood evolution time is considered, which may result in the discharge at the downstream smaller than that at the upstream

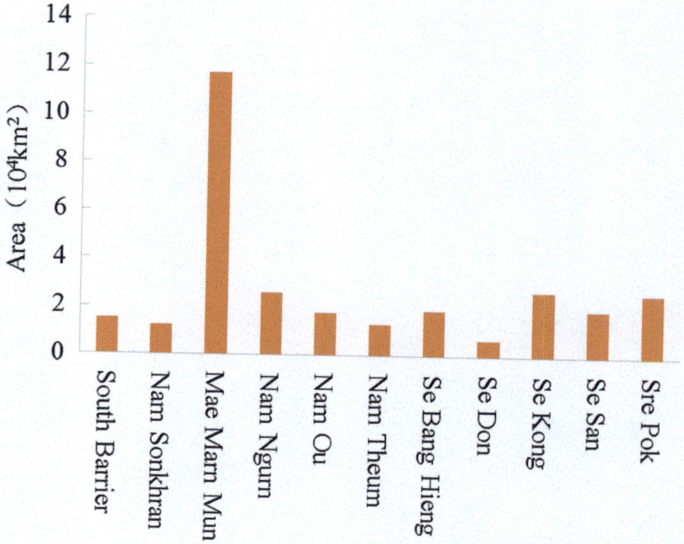

Fig. 3.27 Drainage area of main tributaries

Ms. Hou's research primarily focuses on hydrological processes, hydrological predictions and modelling, and flood management. She is in Department of Hydraulic Engineering of Tsinghua University. She has published over 1 journal papers.

Dr. Liu's research primarily focuses on hydrological processes and modeling, flood management, and transboundary river water cooperation. She has over 5 years' research experiences in transboundary water cooperation of China. She was a participant of various Lancang-Mekong water resources cooperation activities. She is an editor of a joint publication by MWR of China and MRC in 2016. She is Program Managers of 3 international cooperation projects with Mekong River countries. She is author of over 10 journal papers and 1 book.

Dr. Tian's research primarily focuses on hydrological processes and modeling, flood management, agricultural water management, and transboundary river water management. He is editor of two journals of *Hydrology and Earth System Sciences* and *Journal of Hydrology*. He is the Chair of Panta Rhei Initiative (the IAHS Scientific Decade 2013–2022) for the fourth biennium (2019–2020), and also the chair of Biofuel Working Group of ICID from 2015 to the present. He has published over 150 journal papers, 3 books and 1 textbook.

Chapter 4
Analysis of Drought Character in the Mekong River Basin

Xuejun Zhang and Hui Liu

Abstract Drought is among the most costly natural disasters in the Mekong River Basin (MRB). To understand the spatial and temporal characteristics of drought can largely facilitate scientific drought management and risk mitigation. Using a set of long-term (1901–2016) global monthly precipitation data and nearly 30-year (1985–2016) *in situ* daily observed streamflow, this chapter investigates the long-term trend and inter-annual variability of meteorological and hydrological droughts in MRB, respectively, through estimations of multi-scale Standardized Precipitation Index (SPI) and Standardized Runoff Index (SRI). Results indicate that while with slight upward trend from basin-scale perspective, the SPI is found with obvious downward trend over the northeastern Thailand, most of Cambodia and Myanmar, suggesting these regions are overall subject to intensified drying during the past half-century. The occurrence frequency of drought is over 25% across much of MRB, particularly for southern Cambodia and Mekong delta where the occurrence of extreme drought is around 10%. The hydrological drought analysis show that the trend of SRI greatly varies by *in situ* station location over the past 30 years. The 12-month SRI (SRI-12) for the Upper station (Chiang Saen) exhibits obvious decreasing trend, and mostly falls into the negative range, suggesting the recent frequent below-normal streamflow in Chiang Saen station. The occurrence of hydrological drought in the Middle portion is found with downward trend, as indicated with the long-term increasing SRI-12 at NongKhai and Mukdahan stations. When it comes to the downstream station (Stung Treng), while the trend of SRI seems little change, the SRI-12 value is mostly below zero since 2005, suggesting that this portion is susceptible to hydrological droughts recently. These analysis provide us an overview insight into changes in meteorological and hydrological droughts, and advance our understanding on drought variations and its long-term trend in MRB.

X. Zhang (✉) · H. Liu
China Institute of Water Resources and Hydropower Research, A-1 Fuxing Road, Haidian District, Beijing 100038, China
e-mail: zhangxj@iwhr.com

H. Liu
e-mail: wuyouliuhui@163.com

H. Liu (ed.), *Flood Prevention and Drought Relief in Mekong River Basin*, Springer Tracts in Civil Engineering,
https://doi.org/10.1007/978-981-15-2006-8_4

4.1 Data and Methodology

4.1.1 Data

(1) Historical long-sequence rainfall data of the Mekong River Basin in the recent 100 years (1901–2016) was collected and compiled on the basis of the CRU (Climatic Research Unit) global precipitation product. It is a full-coverage, high-resolution and no-omission data set of average monthly surface climatic elements rebuilt by the U.K. University of East Anglia's Climatic Research Unit (CRU) through integrating existing famous databases.[1] The current sequence length is 1901–2016, spatial resolution is 0.5° × 0.5°, and all land networks in the world are covered.
(2) The historical long-sequence section flow data of major hydrological stations of Chiang Saen, Luang Prabang, Nong Khai, Nakhon Phanom, Mukdahan, Pakse, Stung Treng, Kratie on mainstream Mekong River (from upper to lower reaches, see Fig. 3.1 and Table 3.1) was collected from Mekong River Commission. The sequence length is 1985–2016, and temporal resolution is daily-scale observation.

4.1.2 Methodology

In this project, we established the standardized precipitation index (SPI) and standardized runoff index (SRI) as indicators for monitoring and diagnosis of different types of drought and analyze the characteristics of drought, on different scales and of different types, in the Mekong River, from meteorological and hydrological perspectives.

(1) **Definition and calculation of SPI**

Generally speaking, precipitation abides by skewed distribution rather than normal distribution. In drought monitoring and evaluation, Γ distribution probability is usually adopted to describe the variation of precipitation. The standardized precipitation index (SPI), for measuring the excess and deficit of precipitation on various temporal scales, is a widely adopted index for drought diagnosis. Γ distribution probability is adopted to describe precipitation in the SPI calculation; then, normal standardization of skewed probability distribution is conducted; finally, drought is graded using the distribution of cumulative frequency of standardized precipitation. The SPI is an indicator expressing the precipitation occurrence probability in a given period that is applicable to meteorological drought monitoring and evaluation on or above the monthly scale. With the advantages of easy access to data, easy calculation, flexi-

[1]Harris I, Jones PD, Osborn TJ, et al. Updated high-resolution grids of monthly climatic observations-the CRU TS3.10 dataset. International Journal of Climatology, 2014, 34(3): 623–642.

Table 4.1 SPI-based drought classification

Grade	Type	SPI
I	No drought	>− 0.5
II	Mild drought	(−1.0, −0.5)
III	Moderate drought	(−1.5, −1.0)
IV	Severe drought	(−2.0, −1.5)
V	Exceptional drought	≤− 2.0

ble temporal scale and regional comparability, SPI has been widely applied to the depiction of meteorological drought in recent years. SPI formula is[2]:

$$SPI = S\left(t - \frac{(c_2 t + c_1)t + c_0}{[(d_3 t + d_2)t + d_1]t + 1.0}\right) \tag{4.1}$$

$$t = \sqrt{\ln \frac{1}{G(x)^2}} \tag{4.2}$$

In specific, x is precipitation sample value; S is the positive and negative coefficients of probability density; c_0, c_1, c_2 and d_1, d_2, d_3 are calculation parameters of the simplified approximation analysis formula for converting Γ distribution probability into cumulative frequency—$c_0 = 2.515517$, $c_1 = 0.802853$, $c_2 = 0.010328$, $d_1 = 1.432788$, $d_2 = 0.189269$ and $d_3 = 0.001308$. $G(x)$ is rainfall distribution probability related to Γ function. According to the probability density integral formula of Γ function is:

$$G(x) = \frac{2}{\beta^{\gamma} \Gamma(\gamma_0)} \int_0^x x^{\gamma - 1} e^{-x/\beta} dx, \quad x > 0 \tag{4.3}$$

When $G(x) > 0.5$, $S = 1$; when $G(x) \leq 0.5$, $S = -1$.

Drought is graded with SPI ≤ − 0.5 being the standard for judging the occurrence of drought. Table 4.1 shows the drought gradation based on SPI.

(2) **Definition and calculation of SRI**

By reference to the calculation principle of SPI, the Standardized Runoff Index (SRI) was proposed. Based on long-sequence measurement or simulated monthly runoff calculation, SRI measures effectively runoff deficit relative to multi-year average runoff, expresses the probability of occurrence of the cross-section runoff of a given period in the same period in history, and is used in hydrological drought diagnosis and evaluation on and above monthly scale.

---

[2]McKee T, Doesken N, Kleist J. The relationship of drought frequency and duration to time scales. In Proceedings of the 8th Conference on Applied Climatology, Anaheim, CA, USA, 17–22 January 1993; American Meteorological Society: Boston, MA, USA, 1993; pp 179–184.

Similar with the SPI-based gradation of drought, drought is also graded with SRI ≤ -0.5 being the standard of judging hydrological drought, as shown in Table 4.1.

4.2 Character Analysis of Drought

4.2.1 Character of Meteorological Drought

Based on the long-sequence monthly precipitation data for 1901–2016 from CRU, The one-month, three-month, six-month and twelve-month standardize precipitation indexes (SPI1, SPI3, SPI6 and SPI12) was respectively calculated to depict the short-term and long-term meteorological drought in the Mekong River Basin. Based on the SPI calculation results on various temporal scales, we revealed the spatio-temporal characteristics of meteorological drought in the Mekong River Basin from the perspectives of drought severity and drought frequency.

(1) **Drought severity**

1) **Inter-annual variation of SPI**

Figure 4.1 demonstrates the inter-annual variation characteristics of SPI on different temporal scales between 1950 and 2016. In specific, Fig. 4.1a–d show the one-month, three-month, six-month and twelve-month SPI values. Different temporal scales demonstrate great differences in the degree of drought. For instance, SPI1 of the Mekong River Basin in March, 2013 is −1.80, indicating severe drought; SPI3 is −1.43, indicating moderate drought; SPI6 is −0.76, indicating mild drought, and SPI12 is 0.48, indicating no drought. This means the mild or moderate drought on

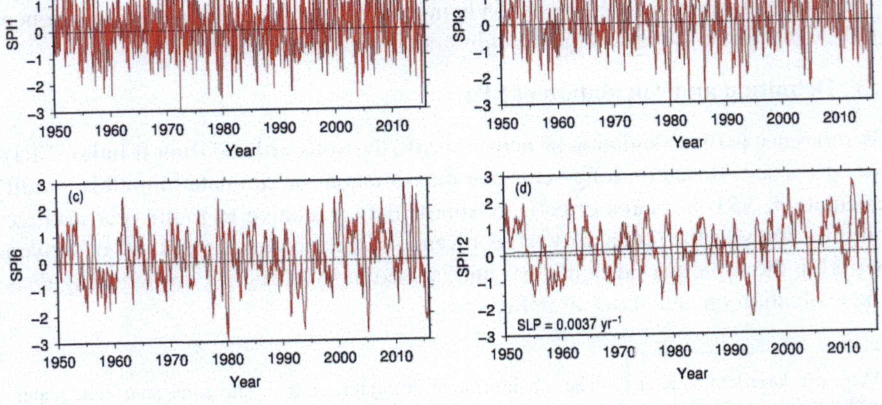

Fig. 4.1 SPI Sequences on Different Temporal Scales of the Mekong River Basin. **a** SPI1; **b** 3 SPI3; **c** SPI6; **d** SPI12

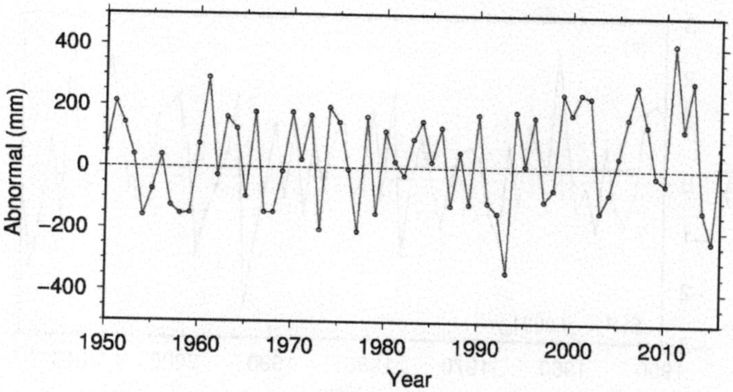

Fig. 4.2 Annual precipitation anomaly of the Mekong River Basin (mm)

short-time scales may be changed into no drought on long-time scales if rainfall deficit can be remitted in time. On the contrary, the severe or exceptional drought on long-time scales may be also no drought or mild drought on short-time scales. An example is July, 2005. SPI1 and SPI3 are 0.87 and 0.18, indicating no drought; along with the growth of temporal scale, however, SPI12 is −1.94, indicating severe drought relative to the same period in history. From Fig. 4.1d, we could see that SPI varies from −2.38 to 2.70, with a slightly upward trend (0.037/10a), meaning the drought severity is slightly decreasing. Based on SPI values on various temporal scales and the gradation standard in Table 4.1, we can know in the past half century, serious droughts gather in late 1950s, the end of 1970s/the beginning of 1980s, the beginning of 1990s and around 2005. On yearly scale, serious droughts took place in 1980, 1992, 1994, 1998, 2005, 2009–2010 and 2015–2016, basically consistent with existing research findings.

Figure 4.2 shows the annual precipitation anomaly of the Mekong River Basin in the past century. Since the 1950s, obvious rainfall deficit was seen in late 1950s, the end of 1970s/the beginning of 1980s and the beginning of 1990s, which further proves the result SPI reveals (Fig. 4.1).

2) **Seasonal variation of SPI**

The Mekong River Basin belongs to the tropical monsoon climate zone, with distinct dry and wet seasonal variations throughout the year: wet season with abundant precipitation from May to October and dry season with low precipitation from November to April of the following year. Therefore, based on the six-month SPI values (SPI6) from October to April, the severity of meteorological drought in rainy and dry seasons of the basin was analyzed.

Figure 4.3a, b show the severity changes of drought in wet season (May–October) and dry season (November-April) respectively in the Mekong River Basin. It can be seen from Fig. that drought and flood events occur alternately in the wet season and the SPI values are between −2.38 and 2.90. On the whole, the variation trend of severity of drought in wet season is not obvious. In terms of inter-annual variations, the wet

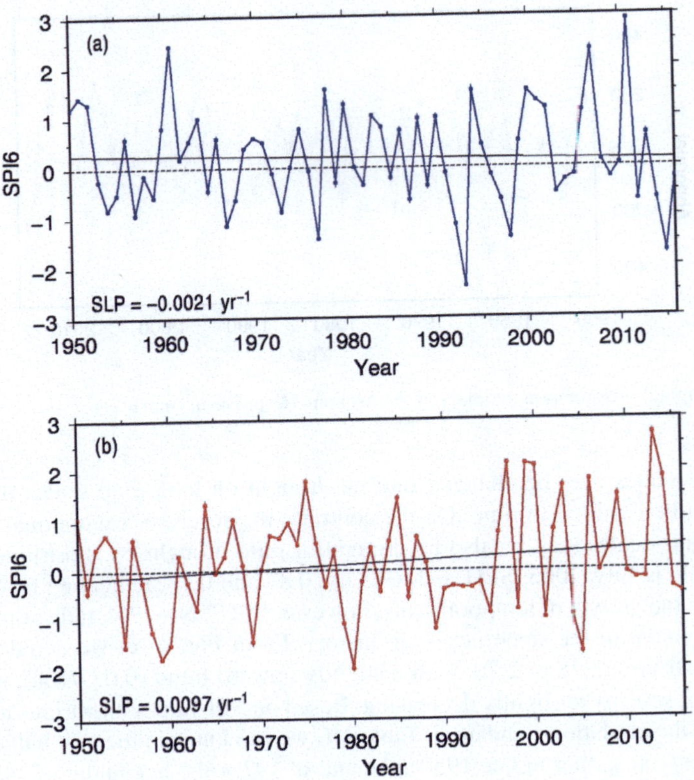

Fig. 4.3 Variation characters of severity of seasonal drought (SPI6) in the Mekong River Basin with the Time. **a** Wet Season (May–October); **b** Dry Season (November–April of Next Year)

season precipitation in the 1960s–1990s and 2005–2010 was relatively high; and the precipitation in the 1950s, 1990s and the last five years (2005–2010) was relatively low. Among them, the wet seasons of 1993, 1998 and 2015 had serious droughts (SPI6 < − 1.5); especially in 1993, SPI6 was only −2.38, indicating exceptional drought.

Compared with the wet season, SPI6 value of dry season in the basin shows an obvious upward trend, indicating that the drought severity in the dry season shows a downward trend. In terms of inter-annual variation, the dry season had less precipitation before the mid-1990s and SPI6 values of 1995, 1959, 1980 and 2005 were lower than −1.5, which was a extraordinarily serious drought year. Since the late 1990s, SPI6 value has been greater than 0 in most years, especially in the dry season of 2013, the SPI6 value reached 2.70 and precipitation was abundant.

3) **Spatial distribution of SPI trend**

Figure 4.4 shows the variation trends of drought severity in the whole year, wet season and dry season of the Mekong River Basin. On the yearly scale, the SPI of Myanmar,

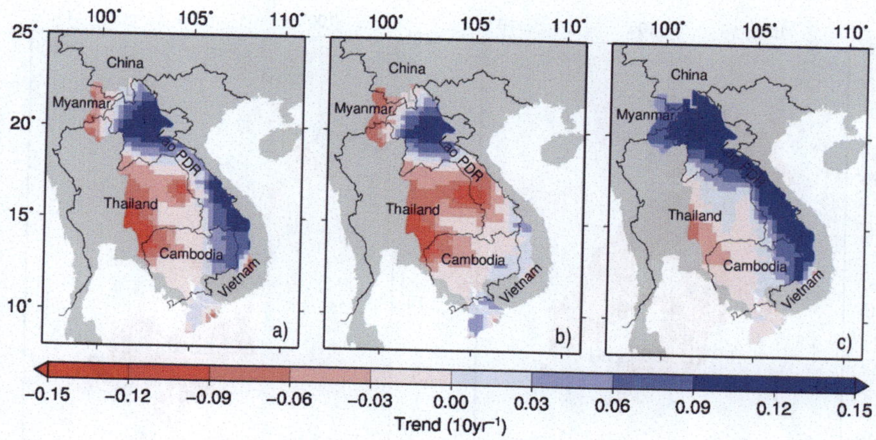

Fig. 4.4 Spatial Distribution of Drought Severity trend in the Mekong River Basin. **a** Annual drought severity trend of SPI (SPI12) on the twelve-month scale; **b** wet season drought severity trend of SPI (SPI6) on the six-month scale; **c** Dry season drought severity trend of SPI (SPI6) on the six-month scale

northeastern Thailand and most parts of Cambodia shows a downward trend, indicating that the annual drought severity in the above areas is increasing, especially in parts of northeastern Thailand, the downward trend of SPI reaches −0.15/10 years; relatively speaking, SPI in Lao PDR has a significant increasing trend, indicating the severity of drought in Lao PDR is decreasing. Figure 4.4b shows that the areas with the increased drought severity in wet season are mainly distributed in northeastern Thailand, central Lao PDR, Myanmar, Cambodia and other countries and regions. The drought severity in northern Lao PDR is obviously weakened. Compared with the wet season, the drought severity in dry season shows a weakening trend in most parts of the basin (corresponding SPI value is increasing); relatively speaking, the drought-enhanced areas are mainly concentrated in northeastern Thailand and a small part of western Cambodia.

(2) **Frequency of drought**

The frequency of drought refers to the number of occurences of drought in the entire period of time. The calculation formula is:

$$d = (n/N) \times 100\% \tag{4.4}$$

In specific, n is the number of years in which drought takes place; N is precipitation sequence length (116 years).

Taking SPI < − 0.5 as the criterion of the occurrence of drought, we made statistics of the frequency of drought in each grid of the Mekong River region (Fig. 4.5a). According to the result, the frequency of meteorological drought exceeds 25% in almost all regions across the basin. In specific, drought frequency surpasses 30% in Lao PDR, northeastern Thailand and northwestern Cambodia; relatively speaking,

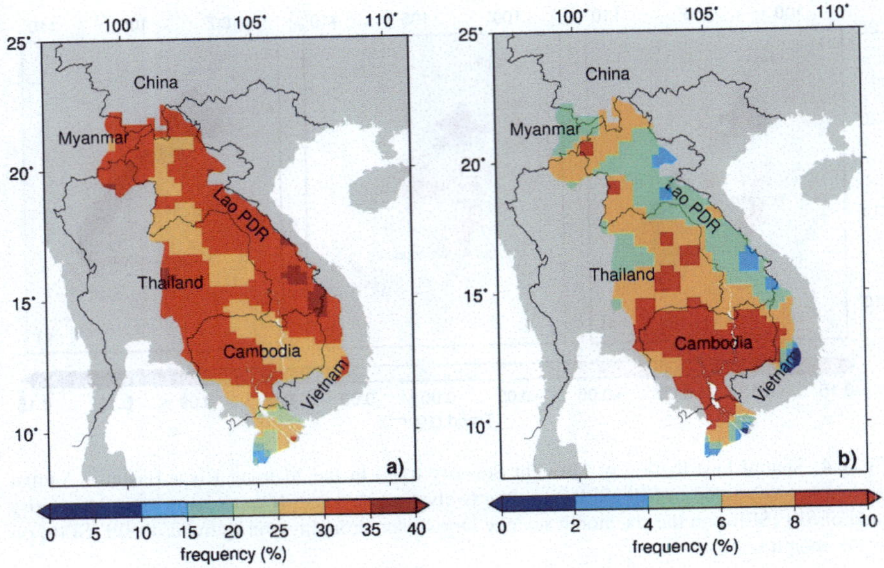

Fig. 4.5 Distribution of annual drought frequency (Left) and frequency of severe and exceptional drought (Right) in the Mekong River Basin

the frequency is slightly lower in Mekong Delta in Viet Nam than in other regions. On this basis, we further calculated the frequency of severe and exceptional droughts (SPI < -1.5) (Fig. 4.5b). According to the result, the frequency of severe and exceptional droughts is high, close to 10%, in Cambodia, Viet Nam and other regions in the lower reaches. Comparison of both figures indicates Mekong Delta registers higher frequency of severe and exceptional droughts than upstream regions, though its drought frequency is slightly lower. This means the Mekong Delta is more liable to severe and exceptional meteorological droughts.

4.2.2 Character of Hydrological Drought

Three typical hydrological stations, Chiang Saen, Mukdahan and Stung Treng, were selected as representatives of hydrological characteristics of upper, middle and lower reaches of the mainstream Mekong River, to analyze the spatio-temporal characteristics of hydrological drought in the mainstream Mekong River. Based on the day-by-day runoffs between January 1, 1985-December 31, 2016, we figured out each station's average monthly runoff in January 1985-December 2016; based on the monthly runoff sequence in the past 32 years (1985–2016), we figured out the one-month, three-month, six-month and twelve-month standardized runoff indexes (SRI1, SRI3, SRI6 and SRI12) respectively for diagnosing each station's hydrological drought on different temporal scales.

Figures 4.6 and 4.7 show SRI1, SRI3, SRI6 and SRI12 of Chiang Saen station, Nongkhai station successively. From Fig. 4.6d we could see that there is a slight downward SRI trend at Chiang Saen station at the rate of about 0.20/10a. Albeit with the little change, the magnitude of SRI is almost negative during 2010–2016, suggesting the frequent hydrological droughts since 2010 in this station. As for inter-annual variation, in the end of the 1980s/the beginning of the 1990s, in 1993/1994,

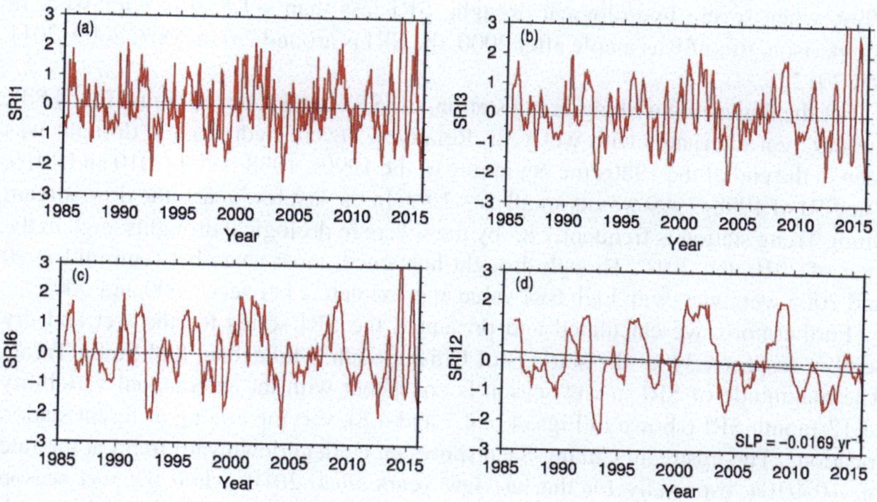

Fig. 4.6 SRI sequence of Chiang Saen station on various temporal scales. **a** SRI1; **b** SRI3; **c** SRI6; **d** SRI12

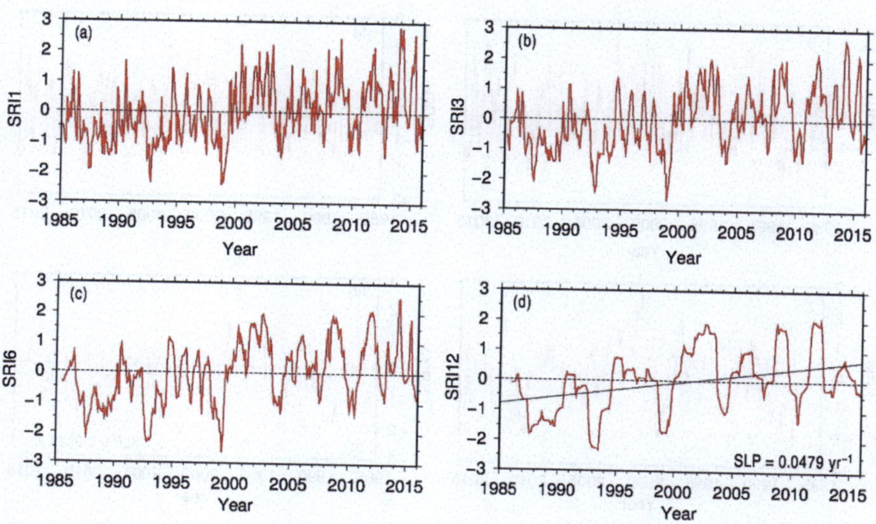

Fig. 4.7 SRI sequence of Mukdahan station on various temporal scales

around 2004/2005 and the last few years since 2010, the streamflow was lower than normal conditions. In particular during 1993/1994, the SRI value is as small as −1.9.

As for the Mukdahan station in the middle reaches, there is an apparent upward SRI trend, with a change rate of 0.05/a (see Fig. 4.7d), which means the hydrological drought severity and frequency both declined. For interannual variation, low runoff gathered in the end of the 1980s/the beginning of the 1990s, 1993–1994 and 1998–1999, when severe hydrological drought, SRI less than −1.5, was witnessed. In comparison, runoff was ample after 2000, the SRI is around 2.0 in 2008, 2009, 2011 and 2012.

For the downstream Stung Treng station, the SRI trend is not obvious (Fig. 4.8d), the dry year alternates with wet year. Relatively severe hydrological drought was seen in the end of the 1980s/the beginning of the 1990s, 1998–1999, 2010 and 2016. The SRI of 1998–1999 was as small as −1.89. In the last few years, the downstream Stung Treng station is frequently hit by the severe hydrological droughts, e.g., in the year of 2010 and 2016. Though drought happened, most years between mid-1990 and 2005 were wet, with high SRI value approximate 2 between 2000 and 2005.

Furthermore, we calculated and presented the SRI series for the wet and dry season, respectively, at the stations of Chiang Saen, Mukdahan, and Stung Treng. The magnitude of SRI in wet season is consistent with the interannual variability of 12-month SRI (shown in Figs. 4.6, 4.7 and 4.8), varying among different station locations. The upstream Chiang Saen shows an evident downward trend at the rate of −0.4/10a, especially for the last few years since 2010, while the wet season SRI in middle Mukdahan station shows the 0.3/10a upward trend. Comparing with the upper two stations, the downstream Stung Treng station seems little change of

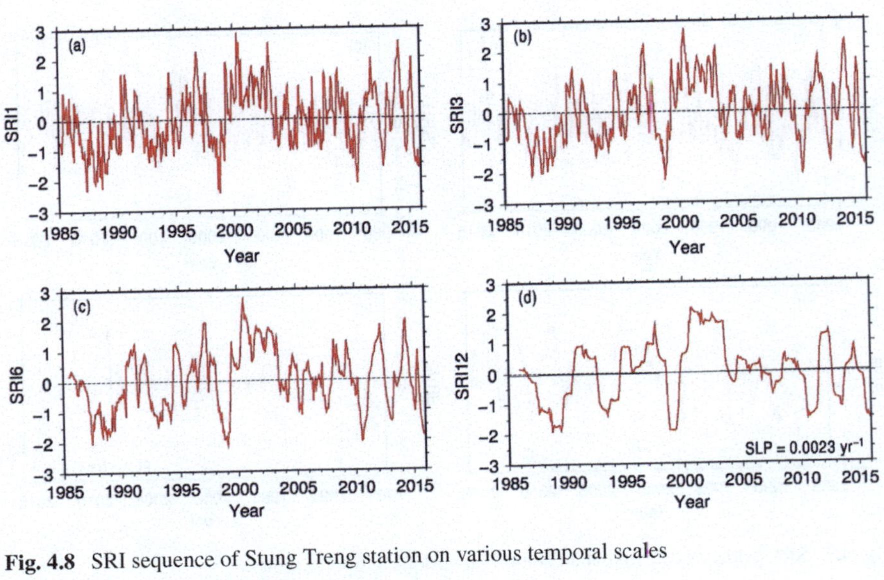

Fig. 4.8 SRI sequence of Stung Treng station on various temporal scales

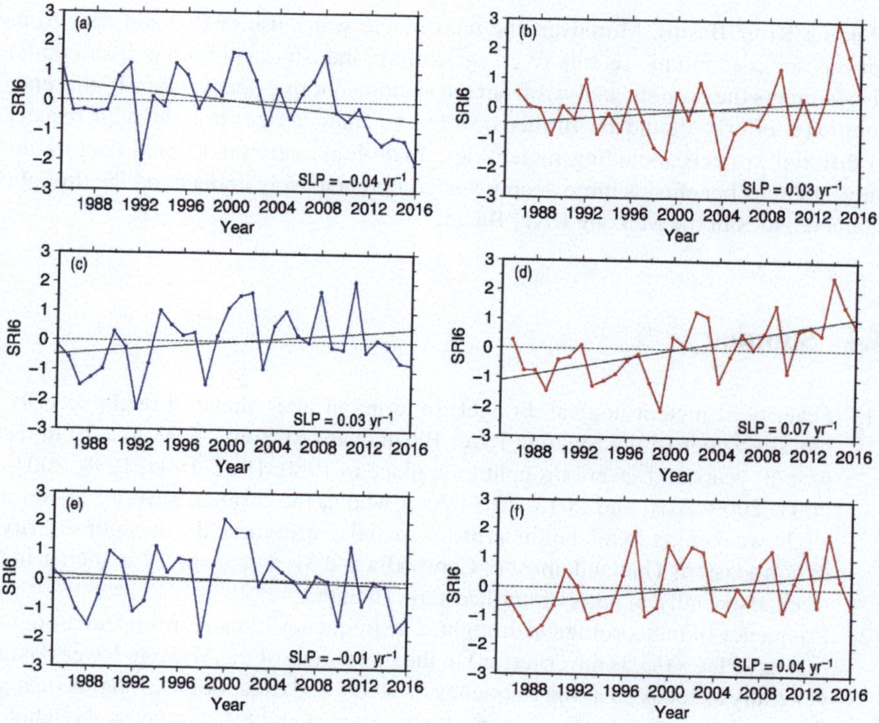

Fig. 4.9 SRI sequence in wet (left panel; May–October) and dry season (right panel; November–April of Next Year) at the Chiang Saen (top panel), Mukdahan (middle panel), and Stung Treng (bottom panel) station

streamflow in wet season. In contrast, all these three typical stations have seen the obvious increasing SRI in dry season with the rate over 0.3/10a, suggesting the overall decreased frequency and severity of hydrological droughts in dry season. This seasonal analysis suggests that more attentions need to be paid on the hydro-meteorological change (e.g., precipitation, streamflow) in wet season in the future (Fig. 4.9).

4.3 Discussion

In this section, the current analysis and results to most extent are limited by the data availability. Specifically, limited by the lack of in situ precipitation measurements across the entire basin, the present analysis of meteorological drought is totally based on the open-accessible global precipitation dataset with relatively low spatial resolution (0.5° × 0.5°). In future, the results may need to be further verified if the abundant ground observations are available (provided by the member counries in

Mekong River Basin). Moreover, the unavailable water use or demand data from various socio-economic sectors (e.g., agriculture, industry, and human livelihoods) also hampers the drought analysis from the socio-economic perspective. To this end, continued efforts should be further devoted to characterize drought with respect to different aspects including meterology, hydrology, agriculture, and socioeconomy, and thus benefits us more deeply and comprehensively understand the drought characteristics in the Mekong River Basin.

4.4 Summary

(1) Severity of meteorological drought: In terms of inter-annual drought severity, the SPI values of the Mekong River Basin changed from −2.38 to 2.70 in the past 60 years and severe drought took place in 1980, 1992, 1994, 1998, 2004–2005, 2009–2010 and 2015–2016. As a whole, the drought severity shows a slight weakening trend, but in terms of spatial distribution, the drought severity in northeastern Thailand, most of Cambodia and Myanmar shows an increasing trend, especially in parts of northeastern Thailand.
(2) Frequency of meteorological drought: The frequency of meteorological drought (using $SPI < -0.5$ as the criterion) in the most areas of the Mekong River Basin generally exceeds 25% and frequency of severe and exceptional droughts (using $SPI < -1.5$ as the criterion) is mostly between 6 and 8%. As far as the whole region is concerned, the areas more liable to severe and exceptional droughts are mostly concentrated in southern Cambodia, Mekong Delta in Viet Nam and other downstream regions.
(3) Severity of hydrological drought: The severe hydrological drought years diagnosed by runoff observations at Chiang Saen station and Mukdahan station in the upper and middle reaches are the late 1980s–early 1990s, 1994, 1998/1999 and 2005; the severe hydrological drought took place in Stung Treng station in the lower reaches in many years, besides the late 1980s–early 1990s and 1998/1999, this station also experienced significant hydrological droughts in 2010 and 2016.
(4) Taking two typical droughts in 2004–2005 and 2016 as examples and combined with meteorological and hydrological factors such as atmospheric circulation background, rainfall and runoff, the causes and development process of two typical droughts were studied. The results show that the El Nino phenomenon, characterized by the atmospheric-ocean circulation anomaly in the tropical Pacific Ocean, usually leads to the early end of the rainy season in the Mekong River Basin and the abnormally less rainfall than that of the same period in history, so that the Mekong River Basin is more liable to a large-scale meteorological drought; and when the meteorological drought lasts for a long time, the corresponding hydrological drought will take place from the upstream to downstream in the basin. Among them, the hydrological drought in the downstream delta

usually leads to the decline of river water level, which makes the time of seawater invasion advanced and the degree enhanced; the superposition of both will aggravate the losses caused by drought, and the drought in the Mekong Delta in 2016 is a typical case.

Dr. Zhang's research primarily focuses on drought monitoring and forecasting, hydrometeorology, and climate change. He has developed an operational drought monitoring and forecasting system over China and the Mekong River basin, respectively, in support of drought early-warning. To date, he has been funded as the director of Chinese Postdoctoral Science Foundation, and participated in more than 10 national-level funding projects. He has published more than 15 papers in the famous high-impact journals like *Nature Climate Change*, and also been involved in 3 published books like the *AGU Geophysical Monograph Series*.

Dr. Liu's research primarily focuses on hydrological processes and modeling, flood management, and transboundary river water cooperation. She has over 5 years' research experiences in transboundary water cooperation of China. She was a participant of various Lancang-Mekong water resources cooperation activities. She is an editor of a joint publication by MWR of China and MRC in 2016. She is Program Managers of 3 international cooperation projects with Mekong River countries. She is author of over 10 journal papers and 1 book.

Chapter 5
Overview of Measures and Assessment of Capacity for Flood Prevention and Drought Relief

Zhixiong Ding, Xuejun Zhang, Song Han, and Hui Liu

Abstract This Chapter introduces the structural and non-structural measures for flood prevention and drought relief in every country of the Mekong River Basin (MRB), and evaluates the capacities for flood prevention and drought relief to these countries. In this chapter, Some specific situations (including location, quantity and scale) of the flood prevention engineering, such as reservoirs, dikes, sluices, pumps, flood storage and detention areas etc., are introduced. The non-structural measures for flood control and disaster reduction of Mekong River Commission and other countries in the basin are introduced from the aspects of hydrological monitoring, prediction, early warning, organization and management system, emergency response, etc. The flood prevention capacity of each country in the MRB and the overall flood control capacity of the MRB are evaluated from the aspects of the per capita reservoir capacity of each country in the basin, and the flood discharge capacity of the main stream of the Mekong River, etc. In this chapter, the drought relief capacity is assessed from an analysis of the available structural and non-structural measures in the MRB using multiple disaster survey data. Specifically, three key indicators (including reservoir irrigation pattern, the proportion of irrigation area, per capita GDP) are employed to assess the water project capacity, economic strength, and drought prevention and disaster reduction response capability. Results show that the flood and drought relief capacity is mostly dependent on the support capacity from the national economy and the water conservancy project. From the national view, the flood and drought

Z. Ding (✉) · S. Han
China Institute of Water Resources and Hydropower Research, Research Center on Flood and Drought Disaster Reduction, A-1 Fuxing Road, Haidian District, Beijing 100038, China
e-mail: dingzx@iwhr.com

S. Han
e-mail: hansong@iwhr.com

X. Zhang · H. Liu
China Institute of Water Resources and Hydropower Research, A-1 Fuxing Road, Haidian District, Beijing 100038, China
e-mail: zhangxj@iwhr.com

H. Liu
e-mail: wuyouliuhui@163.com

H. Liu (ed.), *Flood Prevention and Drought Relief in Mekong River Basin*, Springer Tracts in Civil Engineering,
https://doi.org/10.1007/978-981-15-2006-8_5

relief capacity differs by countries. Comparing with other countries, the Thailand and Viet Nam are of better capacity in response to flood and drought disasters, attributing to the enough economic investment in emergency response system and the relatively completed flood prevention and irrigation project system. In contrast, the flood prevention and irrigation project and its spatial coverage in other countries (Cambodia, Laos and Myanmar) are still in infancy, resulting in the relatively poor flood prevention and drought-coping capacity. From the basin perspective, the integrated flood prevention and drought relief capacity is weak, due to the lack of the Upper-Lower coordination scheme. This suggests the countries in the Mekong River Basin should further strengthen the coordination and communication to improve the integrated water resource emergency management, and thus explore the potential flood prevention and drought relief capacity at the basin-scale.

5.1 Data and Methodology

5.1.1 Data

In order to survey the flood prevention and drought relief measures of the Mekong River Basin countries and evaluate their corresponding capability, the following relevant information was obtained through on-site investigation and data collection.

(1) Maps, photos and videos from the site survey.
(2) Relevant public information downloaded from the Mekong River Commission website, including: incident reports, hydrological sites and their sections, water monitoring data, and mountain flood warning information. Including:

    Planning Atlas of the Lower Mekong River Basin, 2011.
    Overview of the Hydrology of the Mekong Basin, 2005.
    Annual Mekong Flood Report, 2005–2014.

(3) The professional websites of relevant departments of the Mekong River countries and information from other third-party related websites. These web sites are as follows.

    Cambodia National Mekong Committee http://cnmc.gov.kh/cnmc/index.php/en/.
    Laos National Mekong Committee http://www.lnmc.gov.la.
    Thailand National Mekong Committee http://www.tnmc-is.org.
    Viet Nam National Mekong Committee http://www.vnmc.gov.vn/.
    Thailand's Department of Disaster Prevention and Mitigation.
    http://www.disaster.go.th/en/index.php.
    Central Steering Committee for Natural Disaster Prevention and Control of Viet Nam.
    http://phongchongthientai.vn/.

Tonle Sap Lake Authority http://www.tonlesap.gov.kh/.
Asian Disaster Reduction Center http://www.adrc.asia/.

(4) The Google Earth satellite remote sensing image data in recent years in the Mekong River Basin and topographic data of Mekong River Basin with 10 m resolution provided by Google Map.

5.1.2 Methodology

Through on-site investigation, expert consultation, and analysis of relevant data, study and analysis of flood prevention structural measures, drought-resistant structural measures, non-structural measures for flood prevention and drought relief by the Mekong River Basin countries were done to assess the flood prevention and drought relief of these countries. The specific content mainly includes:

- Using the Mekong River Basin Planning Atlas, in combination with on-site investigation, satellite remote sensing image interpretation and other related data analysis, and through the consultation and exchange with relevant experts from the Mekong River countries, the dikes, reservoirs, gate pumps, section of Mekong river mainstream and status, scale, location and related characteristic parameters of flood area in these countries were studied and analyzed.
- Based on the Atlas of Mekong River Basin Planning, the distribution characteristics of irrigation facilities and dam reservoirs in the Basin were studied by means of satellite remote sensing images and field investigations. The analysis is focused on the distribution of reservoirs on the mainstream and tributaries, the distribution of irrigation projects in the basin, the scale of irrigation districts, the proportion of irrigation area, and other hydraulic engineering parameters closely related to drought relief.
- Through the analysis of relevant data, in combination with expert consultation and exchange, and the status of non-structural measures for flood prevention and drought relief in the Mekong River Basin countries was studied, including: the flood prevention and drought relief organization system of Mekong countries, the monitoring, forecast and early warning of Mekong River by the Mekong River Commission and countries in the basin, and situation of flood prevention and drought relief in of the Mekong River Basin countries, etc.
- Based on the analysis of flood prevention structural measures, and non-structural measures in the Mekong River Basin countries, and the further analysis of the flood prevention and disaster reduction organization system, the embankment status, the reservoir storage capacity, the river channel discharge capacity and the flood storage capacity of flood storage areas in these countries, the flood prevention capacity of these countries was analyzed. The flood prevention status of the basin from the perspective of the whole river basin was analyzed, and the experience of basin-level flood prevention management of transnational rivers with the example

of Rhine River Basin was introduced, to provide reference for the implementation of basin-level flood prevention management in the Mekong River Basin.

- Based on structural and non-structural measures for drought relief in the Basin, in combination with economic strength of the Mekong River countries, and taking the size and distribution of the reservoir irrigation area, the proportion of irrigation area, and per capita GDP as indicators, the drought relief capability of these countries was comprehensively investigated and assessed from the aspects of water project capacity, economic strength, and drought prevention and disaster reduction response capability. And furthermore, through the specific case study, the comprehensive drought relief of the Mekong River Basin from a whole basin perspective to tap the greater drought relief potential within the basin was analyzed.

5.2 Structural Measures for Flood Prevention

5.2.1 Cambodia

The flooding in Cambodia is mainly affected by its own topography and meteorological conditions, as well as the upper reaches of the main stream of the Mekong River, and the three rivers of Se Kong, Se San and Sre Pok. In recent years, with the increase of upstream irrigation water consumption and the construction of reservoir projects, the threat of flooding in the floodplain of the Mekong River in Cambodia and the surrounding area of Tonle Sap has been reduced. As a result, the Cambodian government has increased its development in these areas, and these development areas are still threatened by floods. In addition, the reservoir built upstream of the tributary also has the risk of dam break, resulting in flooding in Cambodia. For example, on July 24, 2018, the dam of the Xe Pian-Xe Namnoy reservoir in the province of Api, Laos, caused extensive flooding in the Se Kong river. The dam-breaking flood affected the territory of Cambodia in the lower reaches of the Se Kong River. At the Siem Pang station on the Se Kong River in Cambodia, the water level was 8.30 m at 3 p.m. on July 22, and 12.47 m at 2 p.m. on July 27, an increase of 4.08 m.[1]

(1) **Reservoirs**

Along with economic and social development, Cambodia has built 2 reservoirs, one of which is O Chum 2 with the storage capacity of 120,000 m^3, the other one is Lower Se San 2 with the storage capacity of 2.715 km^3, and plans to build another 12 reservoirs (Fig. 5.1), with a gross storage capacity of around 16 km^3. Among the 12 planned reservoirs, 5 reservoirs are located in the mountainous region on the upper Tonle Sap Lake, with a gross storage capacity of 5.03 km^3; 5 reservoirs are located

[1] MRC monitors the flooding situation in Southern Laos closely, http://www.mrcmekong.org/news-and-events/news/mrc-monitors-the-flooding-situation-in-southern-laos-closely/. 25th July 2018.

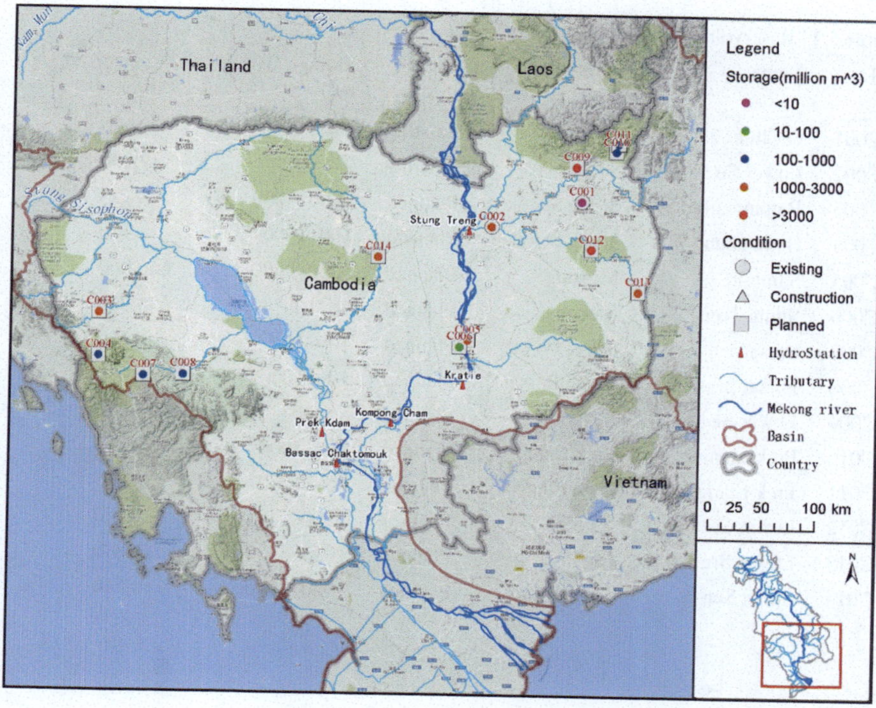

Fig. 5.1 Distribution of reservoirs in Mekong River Basin in Cambodia

in the eastern mountainous region, with a gross storage capacity of about 9 km^3; and the rest 2 reservoirs are located on the mainstream Mekong River, with a gross storage capacity of 2.07 km^3. Informations of the reservoirs are listed in Table 5.1. The distribution see Fig. 5.1 (labeled with ID).

(2) **Levee projects**

In Cambodia, river channels are basically in a natural state without levees or other flood prevention projects on both banks; instead, there are some urban bank protection works in only Stung Treng, Kratie, Kompong Cham, Phnom Penh and other cities. On the right bank of the Mekong River in Cambodia is the catchment basin of the Tonle Sap Lake, which not only takes in, regulates and stores floodwater (about 57% of lake water) of the Mekong River and also collects water (about 30% of lake water) from surrounding tributaries, with a gross catchment area of 85,850 km^2 (including the area of the lake region). Residents inhabit along the lake. About 1.5 million people live on fishing, collection of wild animals and plants in wet land or cultivation of few cereals in floating houses or hamlets near the lake. Near Stung Treng City on the left bank of the Mekong River in Cambodia, three big tributaries flow together into the Mekong River, and they are the Se Kong River that originates from Lao PDR and the Se San River and the Sre Pok River with their sources in Viet Nam. Gross catchment area of the three rivers is 78,645 km^2 or about 13% of the drainage area

Table 5.1 Reservoirs in Cambodia

ID	Name	River	Status	Storage (10^6 m^3)	Drainage area (km^2)
C001	O Chum 2	O Chum	Existing	<10	45
C002	Lower Se San2	Se San	Existing	2715	49,200
C003	Battambang 1	Stung Sangker	Planned	1000–3000	2135
C004	Battambang 2	Stung Sangker	Planned	100–1000	120
C005	Sambor	Mekong	Planned	1000–3000	646,000
C006	Stung Treng	Mekong	Planned	10–100	635,000
C007	Pursat 1	Stung Pursat	Planned	100–1000	1100
C008	Pursat 2	Stung Pursat	Planned	100–1000	2080
C009	Lower Se San 3	Se San	Planned	>3000	15,600
C010	Prek Liang 1	Prek Liang	Planned	100–1000	883
C011	Prek Liang 2	Prek Liang	Planned	100–1000	595
C012	Lower Sre Pok 3	Sre Pok	Planned	>3000	26,200
C013	Lower Sre Pok 4	Sre Pok	Planned	1000–3000	13,800
C014	Stung Sen	Stung Sen	Planned	1000–3000	10,540

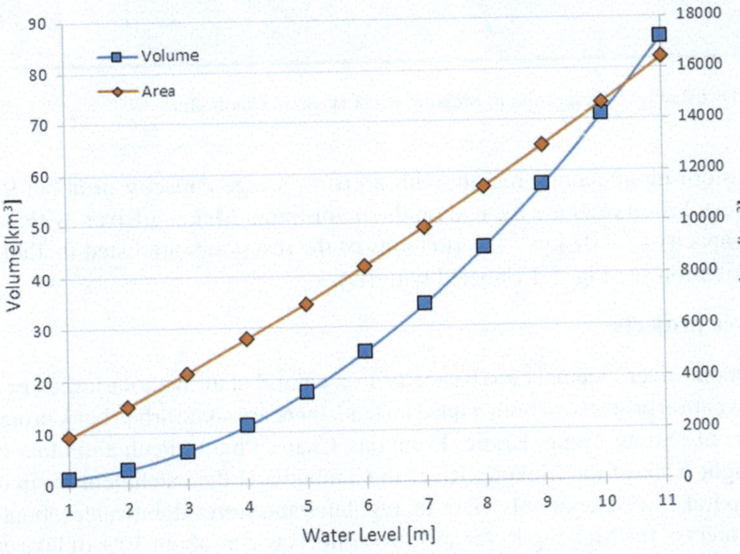

Fig. 5.2 Water level-area-volume curve of Tonle Sap Lake

of the Mekong River. Under the influence of the Annam Mountainous Region, these rivers contribute to about 17% of the average runoff of the basin. Their afflux into the mainstream gathers in August-October. After joining the runoffs from upstream, the water pushes up the flood peak of the mainstream, submerges the flood plains of the Mekong River in Cambodia and further enhances the backflow of the Tonle Sap River, with an important hydrological function in the Tonle Sap Lake's regulation of flood in the mainstream Mekong River. Main hydrological stations on Mekong mainstream in Cambodia include Stung Treng and Kratie. The warning water level and flood water level are 47.49 m and 48.79 m at Stung Treng station; that of Kratie station are 20.92 m and 21.92 m respectively. The Mekong mainstream becomes wider in the Kratie section.

(3) **Flood storage and detention areas**

With a special geographical location, Cambodia becomes an important flood storage and detention region in the Mekong River Basin that plays an important role in reducing the flood peak of the mainstream Mekong River, storing and detaining superfluous floodwater of the Mekong River and relieving flood inundation in the lower reaches. From mid-May to early June, the Lancang-Mekong River Basin will usually enter into the wet season under the influence of the East Asian Monsoon and Southwest Monsoon. In mid-July, the flood of the Lancang River will reach the peak. While a huge amount of floodwater rushing through the Khone Falls into the flat river channels in the lower Mekong River in Kratie, water level of the Chaktomuk intersection between the mainstream Mekong River and the Tonle Sap River will rise rapidly, outflow of the Tonle Sap River will be inverted into inflow, and flood of the Mekong River will flow into the Tonle Sap Lake. Upon mid-August, the East Asian Monsoon of western China weakens, but the growing southwest monsoon brings forth a huge rainfall as blocked by the Annam Mountain; in combination with the Pacific Ocean's tropical storm and typhoon in September, rainfall gets intensified in the Mekong River Basin, bringing more floodwater to the Tonle Sap Lake. In the meantime, floodwater below Kratie also overflows into the wetland region of the flood plain of the Mekong River. In late September, backflow of the Tonle Sap Lake gets to the peak and after the short-term high water level, the Tonle Sap Lake will get back to normal direction of flow into the Mekong River in the beginning of October, along with the weakening of the monsoon. From July to September, the storage and detention of floodwater in the Tonle Sap Lake and the flood plains of Cambodia not only can relieve life and property loss of residents in the lower reaches, but also plays a positive and important role in maintaining the diversity of ecosystems surrounding the Tonle Sap Lake and assuring water replenishment in the dry season in the delta region in the Mekong River. The scale of floodwater detained in the Tonle Sap Lake and the flood plains rests with the intensity of tropical storm, the rainfall in the Annam Mountainous Region and tributary replenishment in Lao PDR.

As for the Tonle Sap Lake, water area increases from 2215 km^2 on average in dry season to 13,258 km^2 in flood season, water depth rises from 1 m to 6–10 m, and water volume increases from 1.5 to 70 km^3. According to MRC's research, 57% of water replenishment of the Tonle Sap Lake is derived from the Mekong River,

30% from tributaries of the lake and 13% from surface rainfall. Annual inflow to the Tonle Sap Lake ranges from 44.1 km^3 (1998, exceptionally dry year) to 106.5 km^3 (2000, wet year), with an average of 79 km^3. After the flood peak, water of Tonle Sap Lake flows back to the Mekong River, with 87% flowing through Tonle Sap River to Mekong River, 1% becoming overland flow and 12% evaporated on lake surface. Outflow ranges between 43.5 km^3 (1998) and 104.8 km^3 (2000), with an average of 78.6 km^3. Figure 5.2 is the water level-area-volume curve of Tonle Sap Lake.

In the flood plains of the mainstream Mekong River in Cambodia, floodwater overflows the bank in flood season, making the river as wide as 50 km at maximum. Usually, overflowing floodwater stays for weeks in the flood plains, which play an important role in storing and detaining the floodwater of the mainstream Mekong River, similar with the impounding regions of Tonle Sap Lake. In wet year (e.g. 2000), the flood plains' water storage and retention is equivalent to that of Tonle Sap Lake. In dry year (e.g. 1998), it is close to one third of the water storage and detention of Tonle Sap Lake. On average, the former is equivalent to 1/2 of the latter.

5.2.2 *Lao PDR*

With the influence of the terrain of Annan Mountain, the basin area of Mekong River in Laos which is on the left bank of the Mekong River accounts for 25% of the total area of the River, but the annual average runoff accounts for 35% of the total. The Nam Tha and Nam Ou rivers between Chiang Saen and Luang Prabang contribute 6%; and there was no large tributary between Luang Prabang and Vientiane in Laos on the left bank of the Mekong River, so the runoff only accounts for 1%; the Nam Ngum, Nam Theun, Nam Hinboun and Se Bang Fai between Vientiane and Savannakhet contribute 22% of the annual average total runoff of the Mekong River; the remaining 6% is contributed by some small tributaries of the Mekong River in Laos below Mukdahan, including the upper reaches of the Se Kong River that flows into Cambodia.

(1) **Reservoirs**

There are 100 reservoirs existing, under construction or planned, with a gross storage capacity of 58.63 km^3, in the Mekong River Basin, including the mainstream Mekong River, in Lao PDR, and their locations are shown in Fig. 5.3. In specific, 14 reservoirs have been constructed, with a gross storage capacity of 11.74 km^3; 27 reservoirs are under construction, with a gross storage capacity of 17.45 km^3; 59 reservoirs are planned, with a gross storage capacity of 29.43 km^3; 9 reservoirs are located on the mainstream Mekong River, with a gross storage capacity of 3.16 km^3, including 2 reservoirs under construction—Xayabury and Don Sahong whose storage capacity is 0.225 km^3 and 0.115 km^3 respectively—and 7 reservoirs planned. Information and distribution of these reservoirs see Table 5.2 and Fig. 5.3. These reservoirs are mainly located in Nam Ou, Nam Kan, Nam Ngum basin and upper Se Kong.

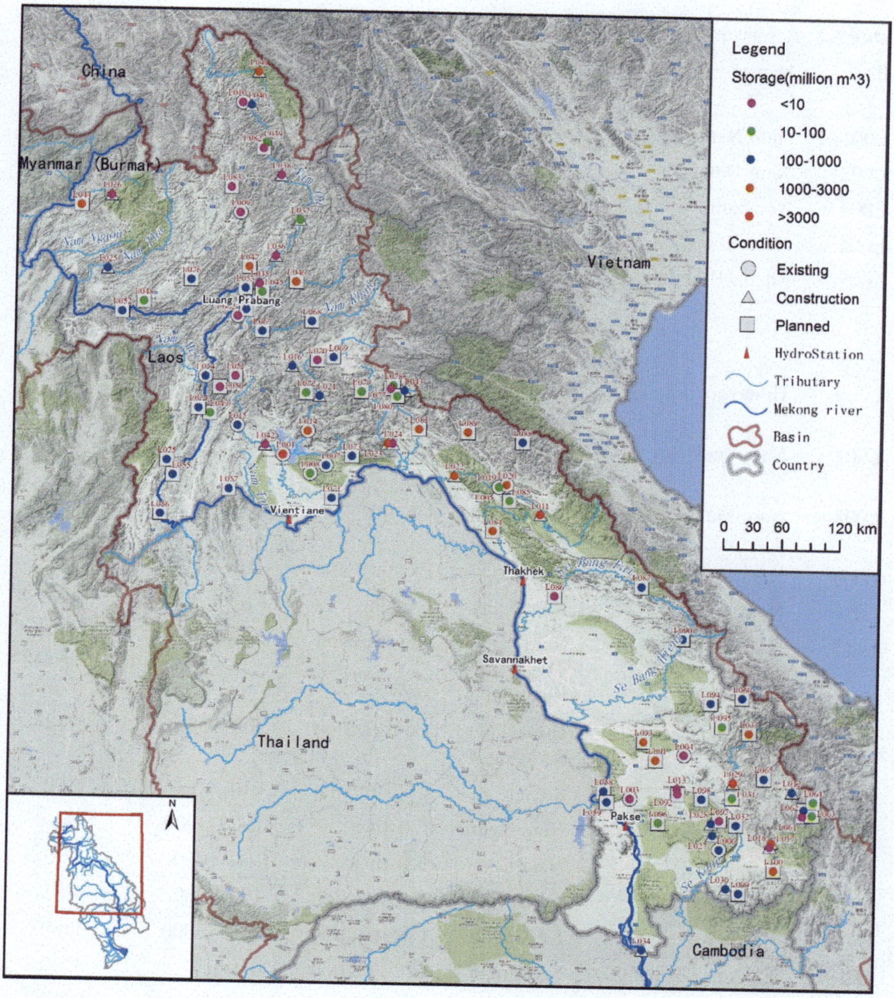

Fig. 5.3 Reservoirs in Mekong River Basin in Lao PDR

(2) Levee projects

Levees on the mainstream Mekong River in Lao PDR include 7.4 km urban revetment in Vientiane, 2.5 km in Pakse and 5 km in Champasak. There are gates and drainage pumping stations for flood prevention in Vientiane, Pakse, Thakhek, Savannakhet, Champasak and other cities, but most reaches along the mainstream Mekong River in Lao PDR only have natural banks, and the estuary of tributaries of Mekong River is basically in a natural state, liable to the influence of high-level flood and backwater effect of Mekong River. Because rainfall is frequent in mountainous areas of Lao PDR, minor basin management is carried out and flash flood warning systems

Table 5.2 Reservoirs in Lao PDR

ID	Name	River	Status	Storage (10^6 m^3)	Drainage area (km^2)
L001	Nam Ngum 1	Nam Ngum	Existing	>3000	8460
L002	Nam Dong	Nam Don	Existing	<10	4
L003	Xelabam	Se Done	Existing	<10	6360
L004	Xeset 1	Se Set	Existing	<10	485
L005	Theun-Hinboun	Nam Theun, Nam Hin Boun	Existing	10–100	8927
L006	Houayho	Houayho, Se Kong	Existing	100–1000	192
L007	Nam Leuk	Nam Leuk, Nam Ngum	Existing	100–1000	274
L008	Nam Mang 3	Nam Mang, Nam Ngum	Existing	10–100	82
L009	Nam Ko	Nam Ko	Existing	<10	223
L010	Nam Ngay	Nam Ngay	Existing	<10	315
L011	Nam Theun 2	Nam Theun, Se Bang Fai	Construction	>3000	4013
L012	Xekaman 3	Houayho, Se Kong	Construction	100–1000	712
L013	Xeset 2	Se Set	Construction	<10	392
L014	Nam Ngum 2	Nam Ngum	Existing	1000–3000	5640
L015	Nam Lik 2	Nam Lik	Existing	100–1000	1993
L016	Nam Ngum 5	Nam Ngum	Construction	100–1000	483
L017	Xekaman 1	Se Kaman	Construction	1000–3000	3580
L018	Xekaman-Sanxay	Se Kaman	Construction	<10	3740
L019	Theun-Hinboun expansion	Nam Theun	Existing	10–100	8937
L020	Theun-Hinboun exp. (NG8)	Nam Theun	Existing	1000–3000	2942
L021	Nam Ngum 3	Nam Ngum	Construction	100–1000	3888
L022	Nam Theun 1	Nam Theun	Construction	1000–3000	14,070
L023	NamNgiep 1	Nam Nhiep	Construction	1000–3000	3700
L024	Nam Ngiep-regulating dam	Nam Nhiep	Construction	<10	3750
L025	Nam Tha 1	Nam Tha	Construction	100–1000	8990
L026	Nam Long	Nam Ma	Construction	<10	156

(continued)

Table 5.2 (continued)

ID	Name	River	Status	Storage (10^6 m^3)	Drainage area (km^2)
L027	Xepian-Xenamnoy	Se Pian, Se Nam Noy	Construction	100–1000	820
L028	Xe Katam	Se Nam Noy	Construction	100–1000	263
L029	Xekong 4	Se Kong	Construction	>3000	5400
L030	Nam Kong 1	Nam Kong	Construction	100–1000	1250
L031	Xe Kong 3up	Se Kong	Planned	10–100	5882
L032	Xe Kong 3d	Se Kong	Planned	100–1000	9700
L033	Xe Kong 5	Se Kong	Planned	1000–3000	2615
L034	Don Sahong	Mekong	Construction	100–1000	553,000
L035	Nam Ou 1	Nam Ou	Construction	10–100	25,979
L036	Nam Ou 2	Nam Ou	Construction	<10	22,568
L037	Nam Ou 3	Nam Ou	Construction	10–100	19,774
L038	Nam Ou 4	Nam Ou	Construction	<10	11,799
L039	Nam Ou 5	Nam Ou	Construction	10–100	10,371
L040	Nam Ou 6	Nam Ou	Construction	100–1000	5527
L041	Nam Ou 7	Nam Ou	Construction	1000–3000	3477
L042	Nam Lik 1	Nam Lik	Construction	<10	5050
L043	Nam San 3	Nam San	Construction	100–1000	155
L044	Nam Pha	Nam Pha	Planned	1000–3000	2837
L045	Nam suang 1	Nam Suang	Planned	10–100	5755
L046	Nam Suang 2	Nam Suang	Planned	1000–3000	5195
L047	Nam Nga	Nam Ou	Planned	1000–3000	2477
L048	Nam Beng	Nam Beng	Planned	10–100	1908
L049	Nam Feuang 1	Nam Feuang	Planned	10–100	714
L050	Nam Feuang 2	Nam Feuang	Planned	<10	374
L051	Nam Feuang 3	Nam Feuang	Planned	<10	184
L052	Pakbeng	Mekong	Planned	100–1000	218,000
L053	Luangprabang	Mekong	Planned	100–1000	230,000
L054	Xayabury	Mekong	Existing	100–1000	272,000
L055	Paklay	Mekong	Planned	100–1000	283,000
L056	Sanakham	Mekong	Planned	100–1000	292,000
L057	Sangthong-Pakchom	Mekong	Planned	100–1000	295,500
L058	Ban Kum	Mekong	Planned	100–1000	418,400

(continued)

Table 5.2 (continued)

ID	Name	River	Status	Storage (10^6 m^3)	Drainage area (km^2)
L059	Latsua	Mekong	Planned	100–1000	550,000
L060	Xe Pon 3	Se Bang Hieng	Planned	100–1000	464
L061	Xe Kaman 2A	Se Kaman	Planned	<10	1970
L062	Xe Kaman 2B	Se Kaman	Planned	100–1000	1740
L063	Xe Kaman 4A	Se Kaman	Planned	10–100	265
L064	Xe Kaman 4B	Se Kaman	Planned	10–100	192
L065	Dak E Mule	Se Kong	Planned	100–1000	127
L066	Nam Khan 1	Nam Khan	Planned	100–1000	7300
L067	Nam Khan 2	Nam Khan	Planned	100–1000	5221
L068	Nam Khan 3	Nam Khan	Planned	100–1000	3392
L069	Nam Ngum 4A	Nam Ngum	Planned	100–1000	1725
L070	Nam Ngum 4B	Nam Ngum	Planned	<10	1818
L071	Nam Ngum, Lower dam	Nam Ngum	Planned	100–1000	16,900
L072	Nam Pay	Nam Ngum	Planned	10–100	157
L073	Nam Mang 1	Nam Mang	Planned	100–1000	577
L074	Nam Pouy	Nam Phuai	Planned	100–1000	1680
L075	Nam Poun	Nam Pung	Planned	100–1000	1700
L076	Nam Ngao	Nam Ou	Planned	100–1000	157
L077	Nam Chian	Nam Nhiep	Planned	<10	271
L078	Nam Ngieu	Nam Nhiep	Planned	10–100	309
L079	Nam Pot	Nam Nhiep	Planned	10–100	60
L080	Nam San 3B	Nam San	Planned	10–100	257
L081	Nam San 2	Nam San	Planned	1000–3000	1429
L082	Nam Pok	Nam Ou	Planned	<10	466
L083	Nam Phak	Nam Ou	Planned	<10	594
L084	Nam Hinboun 1	Nam Hin Boun	Planned	1000–3000	1380
L085	Nam Hinboun 2	Nam Hin Boun	Planned	10–100	31
L086	Xe Bang Fai	Se Bang Fai	Planned	<10	6350
L087	Xe Neua	Se Bang Fai	Planned	100–1000	915
L088	Nam Theun 4	Nam Theun	Planned	100–1000	660
L089	Nam Mouan	Nam Theun	Planned	1000–3000	1652
L090	Xe Bang Hieng 2	Se Bang Hieng	Planned	100–1000	669

(continued)

Table 5.2 (continued)

ID	Name	River	Status	Storage (10^6 m^3)	Drainage area (km^2)
L091	Xedon 2	Se Done	Planned	1000–3000	4090
L092	Xe Set 3	Se Done	Planned	<10	127
L093	Xe Bang Nouan	Se Bang Nouan	Planned	1000–3000	474
L094	Xe Lanong 1	Se Bang Hieng	Planned	100–1000	1415
L095	Xe Lanong 2	Se Bang Hieng	Planned	10–100	339
L096	Nam Phak	Houay Namphak	Planned	10–100	80
L097	Xe Nam Noy 5	Se Kong	Planned	<10	60
L098	Houay Lamphan	Se Kong	Planned	100–1000	140
L099	Nam Kong 2	Se Kong	Planned	100–1000	860
L100	Xe Xou	Se Kong	Planned	1000–3000	1273

are built also in some mountainous regions. Lao PDR has 6 hydrological stations—Luang Prabang, Vientiane, Paksane, Thakhe, Savannakhet and Pakse—on the mainstream Mekong River, and the corresponding locations are demonstrated in Fig. 5.3. The warning water level and flood water level of Luang Prabang is 284.695 m and 285.195 m respectively; that of Vientiane is 168.36 and 170.54 m; that of Paksane is 155.625 and 156.625 m; that of Thakhek is 142.629 m and 143.629 m; that of Savannakhet is 137.022 and 138.022 m; that of Pakse is 97.49 and 98.49 m.

5.2.3 *Myanmar*

The Mekong River Basin in Myanmar is made up of the transboundary river between Lao PDR and Myanmar and some tributaries in mountainous regions, involving a drainage area of 28,600 km^2 and an average annual runoff of 17.63 km^3, and floodwater there is largely sourced from the Lancang River and flash floods of surrounding tributaries (see Fig. 5.4). According to Myanmar experts, as it mainly covers mountainous regions, resident population is small, and river channels along the mainstream Mekong River are in a natural state without such infrastructure as flood prevention projects or such water monitoring facilities as hydrological stations. There are minor basin management projects in mountainous regions that can prevent flash flood to some extent.

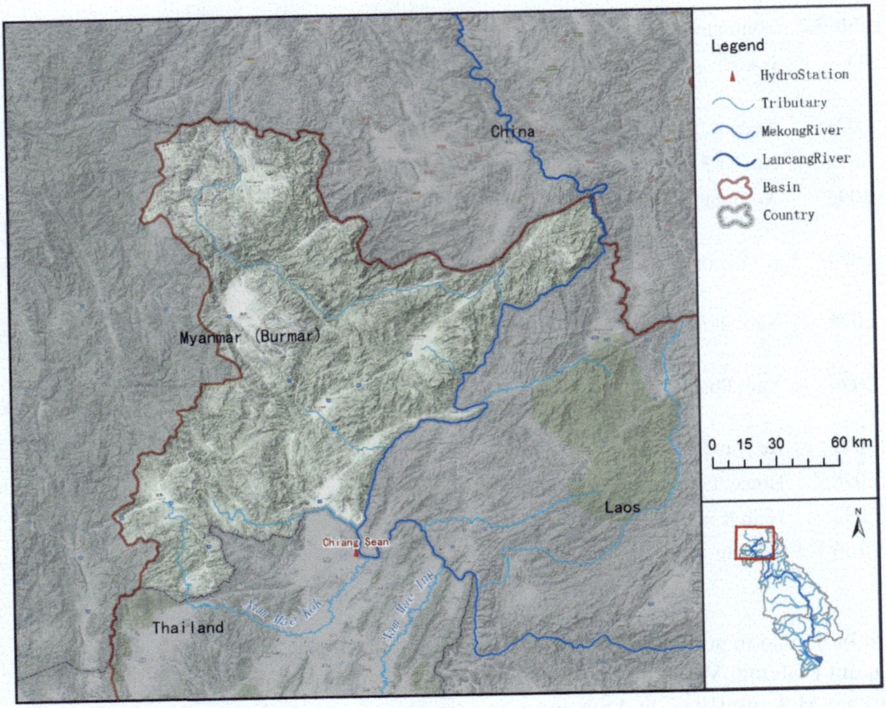

Fig. 5.4 Mekong River Basin in Myanmar

5.2.4 *Thailand*

Thailand's floods in the Mekong River Basin are mainly affected by the high water level of the main stream of the Mekong River, causing flooding in the low-lying areas along the Mekong River or the flooding in the tributaries of Thailand because of the backflow of the high water level of Mekong River. In the northern part of Thailand, it is mainly the two larger tributaries of Mae Nam Kok and Mae Nam Ing; in the dish-like highland area in northeastern Thailand, the main tributaries are Huai Luang, Nam Songkhram and Nam Mun. These tributaries are susceptible to high flood levels in the main stream of the Mekong River without gates and pumps.

(1) **Reservoirs**

There are 7 reservoirs (Fig. 5.5), with a gross storage capacity of 3.57 km^3, in the Mekong River Basin in Thailand. In specific, 7 reservoirs are built up and 1 reservoir is under construction. Information of these reservoirs see Table 5.3 and Fig. 5.5.

(2) **Levee projects**

According to field investigation, consultation and remote sensing interpretation, there are revetment bank along urban river sections on the Thailand side of Mekong River,

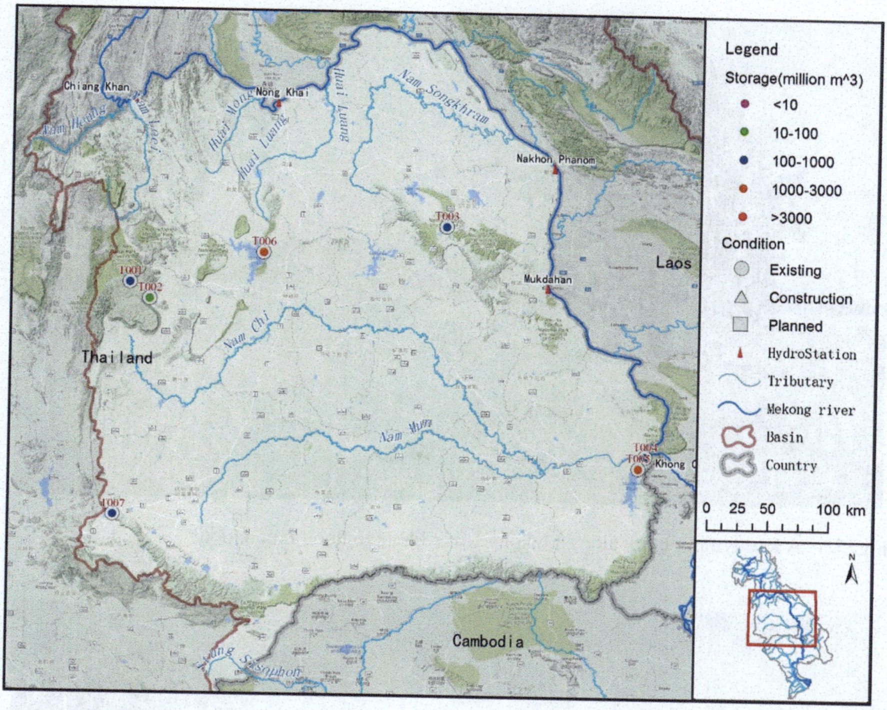

Fig. 5.5 Reservoirs in Mekong River Basin in Thailand

Table 5.3 Reservoirs in Thailand

ID	Name	River	Status	Storage (10^6 m^3)	Drainage area (km^2)
T001	Chulabhorn	Nam Phrom	Existing	100–1000	545
T002	Huai Kum	Nam Phrom	Existing	10–100	282
T003	Nam Pung	Nam Pung	Existing	100–1000	296
T004	Pak Mun	Nam Mun	Existing	100–1000	117,000
T005	Sirindhorn	Lam Dom Noi	Existing	1000–3000	2097
T006	Ubol Ratana	Nam Pong	Existing	1000–3000	12,104
T007	Lam Ta Khong P.S.	Lam Ta Kong	Existing	100–1000	1430

with a total length of around 300 km. Levees formed by road could be found in some low-lying sections. But low-lying sections are still under the impact of the high-level flood of the Mekong River (Fig. 5.6). In some key cities and towns, revetment projects (Fig. 5.7) that can withstand high flood levels are built with high quality.

Fig. 5.6 A low-lying section along Mekong River liable to flood inundation

Fig. 5.7 A revetment project of Mekong River in Thailand

Thailand's hydrological monitoring stations on the mainstream Mekong River are Chiang Saen, Chiang Khan, Nongkhai (see Fig. 5.8), Nakhon Phanom, Mukdahan and Khong Chiam. The warning water level and flood water level of Chiang Saen is 368.61 m and 369.91 m respectively; that of Chiang Khan is 208.618 and 210.118 m; that of Nongkhai is 165.048 and 165.848 m; that of Nakhon Phanom is 142.461 and 143.021 m; that of Mukdahan is 136.219 and 136.719 m; that of Khong Chiam is 102.53 and 103.53 m.

Fig. 5.8 Nong Khai station

(3) **Sluices and pumps**

In the disk-shaped highlands in the northeastern Thailand within the Mekong River Basin, high-level backflow of the Mekong River may be easily caused, flood drainage may be impeded and inland inundation may be incurred where there are no sluices and other flood prevention facilities once sudden rainstorm takes place in the high-water-level season of the Mekong River, though flood inundation is less probable. An example is the Huai Luang regulation project (Fig. 5.9). Irrigation department, responsible for the regulation of the sluice, closes the sluice to raise water level and provide irrigation water in the region when irrigation water is needed; and closes the sluice and discharges floodwater using water pumps to the Mekong River when flood takes place in the Mekong River and water level of the sluice is high (Figs. 5.10 and 5.11). Presently, the project is planned further (Fig. 5.12) to draw water through multi-stage pumping station from Mekong River for irrigation in dry season, and discharge floodwater also through the multi-stage pumping station from the irrigation region to the Mekong River in flood season.

Fig. 5.9 Huai Luang regulation sluice

Fig. 5.10 Flood drainage pump at Huai Luang Sluice

Fig. 5.11 Flood drainage pipe at Huai Luang

Seen from the satellite image, at least 5 tributaries in Khorat highland of northeastern Thailand are installed with sluices, pumps and other projects, and their locations are shown in Fig. 5.13. These sluices aim to provide irrigation water and also prevent flood.

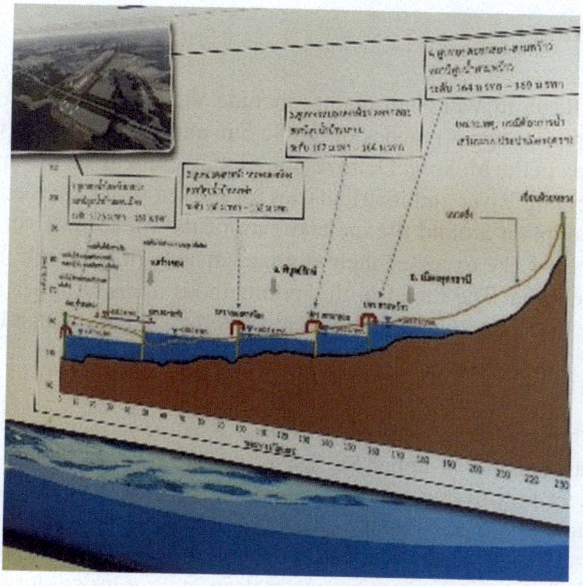

Fig. 5.12 Huai Luang Sluice multi-stage pumping station program

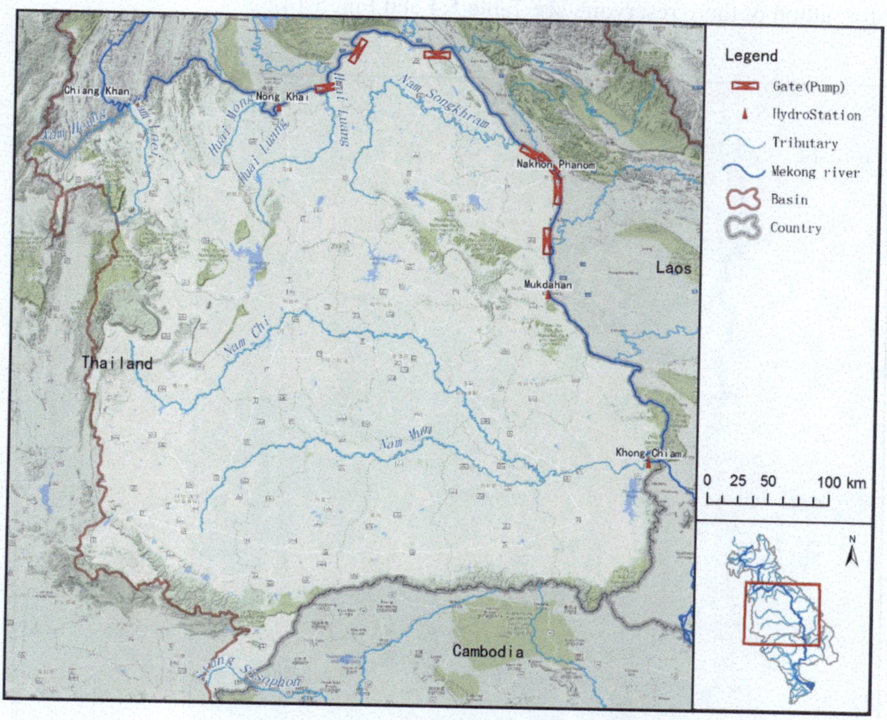

Fig. 5.13 Sluices and gates in Khorat highland of northeastern Thailand

5.2.5 *Viet Nam*

Mekong River Basin in Viet Nam mainly include Viet Nam the Mekong Delta and the mountain areas at upper reaches of the Se San and Sre Pok in the Central Highlands. The floods in the Mekong Delta region mainly come from the upstream of the mainstream Mekong River and its tributaries, or the sea water encroachment. When the flood in floodplain around the main stream of the Mekong River in Cambodia and the Tonle Sap Lake excesses their storage capacity, there might be flooding in the downstream delta. The threat in the mountainous areas at upper reaches of the Se San and Sre Pok in the Central Highlands mainly is the mountain torrent.

(1) **Reservoirs**

In Viet Nam, there are 15 reservoirs, with a gross storage capacity of 3.16 km^3 in the mountainous regions of upper Se San River and Sre Pok River in the Central Highlands. According to our acknowledgement, these reservoirs are designed for power generation and on small and medium scales, and their locations are shown in Fig. 5.14. In specific, 10 reservoirs have been built up, with a gross storage capacity of 2.59 km^3; 3 reservoirs are under construction, with a gross storage capacity of 0.15 km^3; and 2 reservoirs are planned, with a gross storage capacity of 0.41 km^3. Information of there reservoirs see Table 5.4 and Fig. 5.14.

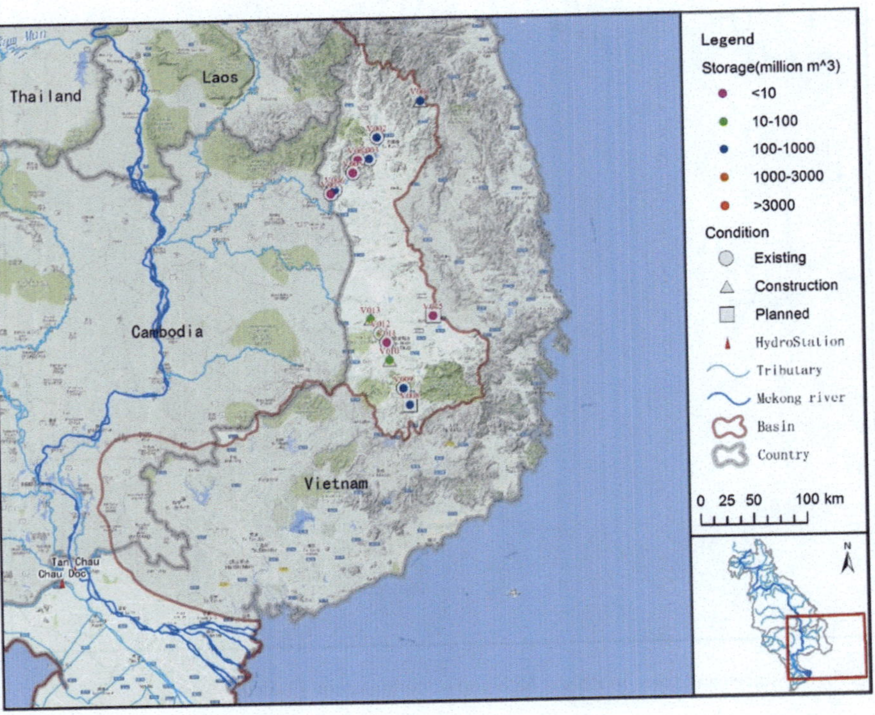

Fig. 5.14 Reservoirs in Mekong River Basin in Viet Nam

Table 5.4 Reservoirs in Viet Nam in Mekong River Basin

ID	Name	River	Status	Storage (10^6 m^3)	Drainage area (km^2)
V001	Upper Kontum	Se San, Dak Bla, Dak Nghe	Construction	100–1000	350
V002	Plei Krong	Se San, Krong Poko	Existing	100–1000	3216
V003	Yali	Se San	Existing	100–1000	7455
V004	Se San 3	Se San	Existing	<10	7788
V005	Se San 3A	Se San	Existing	<10	8084
V006	Se San 4	Se San	Existing	100–1000	9326
V007	Se San 4A	Se San	Existing	<10	9368
V008	Duc Xuyen	Sre Pok, Ea Krong Kno	Planned	100–1000	1100
V009	Buon Tua Srah	Sre Pok, Ea Krong Kno	Existing	100–1000	2930
V010	Buon Kuop	Sre Pok	Construction	10–100	7980
V011	Dray Hlinh 2	Sre Pok	Existing	<10	8880
V012	Sre Pok 3	Sre Pok	Existing	10–100	9410
V013	Sre Pok 4	Sre Pok	Construction	10–100	9568
V014	Dray Hlinh 1	Sre Pok	Existing	<10	8880
V015	Sre Pok 4A	Sre Pok	Planned	<10	9568

(2) **Levee projects**

In the Mekong delta, besides revetments in urban region, most of the river sections do not have high quality protection projects (Fig. 5.15). During the 2017 flood, large-scale bank collapse and erosion took place (Fig. 5.16), in which disaster was avoided through effective prevention and evacuation of people. In the upper Mekong Delta, paddy that relies on fresh water irrigation is the major crop. In dry season, the region is faced with a shortage of fresh water because little water comes from upper Mekong

Fig. 5.15 Natural bank in the Mekong delta

Fig. 5.16 Bank erosion in the Mekong delta

River and intrusion of downstream saltwater. Besides, gates used to block saltwater are mainly located on internal small channels and big channels are not installed with gates, so large-range salinity intrusion takes place during high tide. The lower region is basically exempted from the influence of salinity intrusion because it mainly relies on aquaculture that is resilient to saltwater.

5.2.6 Mainstream Flood Prevention

(1) **From the Chinese border to Chiang Saen, Thailand**

From the Chinese border to Chiang Saen of Thailand, it is mainly the boundary river between Laos, Myanmar and Thailand. This river section is dominated by a mountain valley-type river channel. The river section is trapezoidal. When the water level is low, the river is about 5–6 m deep. In the wet season, the river water level will rise about 10 m, which is almost bankfull. There is a small population around this river section and the runoff mainly comes from Lancang River in China and several small tributaries in Thailand, Laos and Myanmar.

(2) **From Chiang Saen to Vientiane**

The river channel from Chiang Saen to Vientiane becomes broader, and there are terraced river floodplains and crops fields. There are large population-concentrated cities such as Chiang Saen, Chiang Khan and Nongkhai on Thai side; and on Laotian side there are cities with large population such as Luang Prabang and Vientiane. In these urban areas where the population is concentrated, there are generally revetments that can withstand higher water level. The farmland on both sides of the river is mostly natural banks and is vulnerable to flooding. In particular, there are two large tributaries of Mae Nam Kok and Mae Nam Ing along the river section of Chiang Saen on Thai side; and on the Laos side near Luang Prabang, there are several large tributaries such as Nam Ou, Nam Soung, and Nam Khan that originate from Annan mountain area. These tributary floods could easily encounter floods from the Lancang River, resulting in flooding on both sides of the river and causing outer flooding and inner water logging to cities.

(3) Vientiane to the south border of Laos

Vientiane to the south border of Laos, it is mainly boundary river between Thailand and Laos, and the lower half is the inner river section of Laos. This section basically has similar topography of the previous one, and there are more cities with concentrated population on both sides. There are cities such as Nakhon Phanom, Mukdahan and Khong Chiam in Thailand; Thakhek, Savannakhet and Pakse in Laos. In terms of tributary, there are Huai Luang, Nam Songkhram and Nam Mun on the right bank of Thailand and Nam Ngum, Se Bang Fai, Se Bang Hieng and Se Done originating from the Annan mountainous area on the left side of Laos. There are sluices and pumps at the junction of mainstream Mekong River and Huai Luang River on the Thai side. And on the Nam Mun tributary there is a large dam project (Puk Mun Reservoir, with a storage capacity of 125 million m^3). These projects are significant in flood prevention and drainage in the tributary basin on the Thai side. While the tributaries lacking in sluices and dams on Thai side and the incoming tributaries from Laotian side are susceptible to flood from mainstream Mekong River. The floods in farmland and cities on both sides of this river section are mainly from the tributary in the upper reaches and left Annan mountainous areas; and there might be also small amount of flood from tributaries on the right bank in Thailand.

(4) Within Cambodia

The Mekong River is mainly inland rivers in Cambodia. The floodplain of Mekong River and the Tonle Sap Lake in Cambodia stores and detains the flood. Without flooding the downstream, the Mekong River floodplain and Tonle Sap totally stores about 100 km^3 of flood water. The flood storage capacity of the Mekong River floodplain and the Tonle Sap Lake plays an important role in flood regulation and replenishment in Mekong Delta. The Tonle Sap Lake on the right bank not only detains floods in the upper reaches of the Lake District in the mountain area, but also stores floods from the Mekong River during the rainy season and replenishes it to the Mekong River during the dry season. On the left bank there are three major tributaries of Se Kong, Se San, and Sre Pok, which originate from Laos and the Annan Mountains of Viet Nam. The main large population-concentrated cities along the Mekong River in Cambodia include StungTreng, Kratie, Kompong Cham and Phnom Penh. These cities are susceptible to the cumulated impact of floods from the upper reaches of the Mekong River and the three tributaries of Se Kong, Se San and Sre Pok. Except for the cities with concentrated population, it is mostly floodplain on both sides of Mekong River where natural fishery is the major economy and the flood hazard is not serious. But now some areas have been cultivated. Once there is a major flood, the damage will be severe.

(5) Delta

The delta of the Mekong River is mainly in Viet Nam. From the perspective of the basin, this area is most vulnerable to flooding. But with the immense storage and detention by upstream floodplain and Tonle Sap Lake in neighboring Cambodia, it can withstand most floods provide a steady supply of fresh water during the dry season. However, with the development of floodplains in Cambodia, this function may be affected to some extent. And the backflow caused by high water lever could worsen the impact. The delta is mainly dominated by agriculture, and near the coastal areas the fishery. The cities with large populations are Chau Doc, Tan Chau, Long Xuyen, Cao Lamh, Cao Tho and Vinh Long, etc. These cities are all located on the banks of mainstream Mekong River and Bassac River. In the event of a major flood or a high tide, the entire Mekong Delta may be flooded.

(6) **Discharge Capacity of Sections on Mainstream Mekong River**

The Mekong River Commission has played an important role in monitoring, forecasting and early warning of water conditions in the Mekong River Basin. Table 5.5 shows the discharge capability of the main hydrological monitoring stations on mainstream Mekong River based on the statistics of the Mekong River Commission. The flood discharge is calculated according to the relationship between water level and discharge volume based on the cross section at each station (see Sect. 5.2). The capacity of reservoirs built or under construction along Mekong River in Thailand, Laos, Cambodia and Viet Nam is 34.35 km^3, accounting for 8.59% of the Mekong River's average annual total runoff (400 km^3). The planned reservoirs capacity of Laos is 29.65 km^3, that of Cambodia 18.89 km^3, and Viet Nam 0.54 km^3. After all these planned reservoirs have been built, the total reservoir capacity on Mekong River is 83.43 km^3, accounting for 20.86% of Mekong River's average annual total runoff. Under the current conditions, even though these reservoirs are full, the Mekong River itself has more than 90% of runoff floods on a multi-year average; under the planning conditions, it has nearly 80% of runoff floods on a multi-year average. In addition, most of the reservoirs on Mekong River are mainly for power generation and irrigation, and have not been considered for their flood prevention functions. Therefore, the flood prevention situation in the countries along Mekong River has not changed fundamentally.

5.3 Structural Measures for Drought Relief

Improving the water resources support capacity during drought through building irrigation systems like reservoirs, sluices, canals and increasing area of irrigation regions is a major structural measure for drought relief in Mekong River Basin countries. The data used in this section was based on multi-source materials, including MRC publications, web-site information, as well as on-site investigations.

As far as the entire river basin is concerned, nearly 6755 irrigation projects have been put into use, covering an area of 4.78 million hm^2. In terms of spatial distribution, the irrigation area is mainly in the Delta of Mekong River, the northern and northeastern parts of Thailand, the southern part of Cambodia and the upper reaches of the Sre Pok basin in Viet Nam. In terms of seasonal distribution, current irrigation projects are mainly used to cope with the drought that occurs during the wet season rice planting. Specifically, wet-season rice accounts for the largest area of irrigation,

Table 5.5 Discharge capacity of main hydrological monitoring stations on mainstream Mekong River (The correlation between water lever and runoff below Kratie is unstable, and the corresponding runoff of certain flood water level is unavailable)

No.	Name of station	Reach	Country	Warning water level (m)	Flood water level (m)	Flood discharge (m^3/s)
1	Chiang Saen	Upper Mekong River	Thailand	11.5	12.8	17,610
2	Luang Prabang	Upper Mekong River	Lao PDR	17.5	18.0	18,885
3	Chiang Khan	Upper Mekong River	Thailand	14.5	16.0	21,497
4	Vientiane	Middle Mekong River	Lao PDR	11.5	12.5	22,772
5	Nongkhai	Middle Mekong River	Thailand	11.4	12.2	18,658
6	Paksane	Middle Mekong River	Lao PDR	13.5	14.5	23,800
7	Nakhon Phanom	Middle Mekong River	Thailand	11.5	12.0	30,470
8	Thakhek	Middle Mekong River	Lao PDR	13.0	14.0	38,521
9	Mukdahan	Middle Mekong River	Thailand	12.0	12.5	31,811
10	Savannakhet	Middle Mekong River	Lao PDR	12.0	13.0	37,610
11	Khong Chiam	Middle Mekong River	Thailand	13.5	14.5	33,475
12	Pakse	Middle Mekong River	Lao PDR	11.0	12.0	38,899
13	StungTreng	Lower Mekong River	Cambodia	10.7	12.0	66,018

(continued)

Table 5.5 (continued)

No.	Name of station	Reach	Country	Warning water level (m)	Flood water level (m)	Flood discharge (m^3/s)
14	Kratie	Lower Mekong River	Cambodia	22.0	23.0	58,540
15	Kompong Cham	Lower Mekong River	Cambodia	15.2	16.2	–
16	Chaktomuk	Bassac River	Cambodia	12.0	15.0	–
17	Prek Kdam	Tonle Sap River	Cambodia	9.5	10.0	–
18	Tan Chau	Lower Mekong River	Viet Nam	3.5	4.5	–
19	Chau Doc	Bassac River	Viet Nam	3.0	4.0	–

up to 87%; dry-season rice accounts for 31%; third-season rice, which is mainly grown in southern Viet Nam and a small part of Cambodia, accounts for about 37%; the non-rice fields account for about 15% of the irrigated area. At present, there are over 1300 dams/reservoirs for irrigation in the basin, of which most are Earth dam. About half of the dams/reservoirs are located in the Highlands of Viet Nam.

The following is an analysis of the specific structural measures of countries in the basin.

5.3.1 *Cambodia*

Cambodia has built up 2047 irrigation projects, with a gross irrigation area of 479,762 ha in wet and dry season, and average irrigation area per project is 234 ha. In specific, up to 2011, the number of large-scale systems whose irrigation area exceeds 5000 ha amounts to 33, the number of medium-scale systems whose irrigation area ranges between 200 and 5000 ha is 955, and others are small-scale systems whose irrigation area is smaller than 200 ha. In terms of spatial distribution, irrigation structural measures are mainly located in the southern region, especially in Phnom Penh and surrounding Tonle Sap Lake, and their irrigation area takes up 10–30% of local area (district/county), which indicates water use of above regions in dry period can be guaranteed effectively. Relatively speaking, there is almost no irrigation project in other regions, especially in the northeastern region, northwestern region and other remote regions. From the temporal perspective, Cambodia has three harvests of paddy a year. Current projects mainly provide irrigation water for paddy in dry season and in wet season (either around 50%), and little irrigation water is provided to the third-season paddy (between dry season and wet season) and other crops (except paddy). In future 10 years, Cambodia plans to increase nationwide irrigation area to 774,000 ha by 2030.

5.3.2 Lao PDR

Lao PDR has built up 3094 irrigation projects, with an irrigation area of 225,446 ha in wet season. Because most irrigation measures are of small scale, average irrigation area per project in Lao PDR is merely 113 ha. The irrigation area in Lao PDR is much smaller than that in other countries. On the one hand, land suitable for irrigation is limited because of the long, narrow and steep terrain; on the other hand, irrigation input was small in history. In terms of spatial distribution, existing irrigation projects gather in the corridor zones along tributaries of Mekong River and lower flood plains, and mainly aim to irrigate paddy fields. For instance, pumping stations, gates for flow control and 16-km-long conveyance canals are built in Vientiane, Pakse, Thakhek, Savannakhet, Champasak and other cities along the Mekong River. In Lao PDR, irrigation area takes up less than 3% of local area (district/county) and even less than 1% in most regions. Nevertheless, Lao PDR is exerting efforts to develop its hydropower potential and improve water resources project guarantee capacity. On the current basis, Lao PDR plans to build more irrigation projects for dry season rice and non-rice crops irrigation on the plateau region and increase national irrigation area to 213,000 ha by 2030, which will promote engineering support capability for drought relief.

5.3.3 Myanmar

Myanmar is the largest country in the Mekong River, but it is less dependent on the Mekong River because only 2% of its territorial area is located here. Though the area situated in the Mekong Basin is remote region without development, structural measures in other regions of Myanmar is not underdeveloped. In specific, famous dams include North Nawin, South Nawin, Phyu Chaung and Wegyi. In addition, Myanmar has constructed 207 irrigation pipelines covering an area of 250,000 ha. Construction and management of these water resources projects provides effective water resources guarantee and policy support for coping with drought disasters.

5.3.4 *Thailand*

About 36% of Thailand's territorial area, covering 166,000 km^2 and mainly located in the northeast, fall within the Mekong River Basin. On Mekong tributaries in Thailand, 256 dams/reservoirs have been built up for irrigation, with accessory constructions of sluices, pumps and canals, forming a complete set of water division-irrigation system. Take Huai Luang irrigation project as example (Fig. 5.17). It was buit up in 1994, with two-way sluice of three 11 m-wide channels, the maximum drainage capacity is about 990 m^3/s. The irrigation area around the project is about 92,911 ha, of which the low-lying area is 42,227 ha, accounting for about 45%; the remaining 55% (50,684 ha) is the high land. The main crop type there is rice, with an area of 2.02 million ha, makes up 74% of the total area. The remaining 26% is covered by various commercial crops. Along the canals, there are sluices and gates, water will be pumped from mainstream Mekong to the reservoir when drought happens, and then gate will be closed to raise the water level for irrigation in this region. Thailand has built up 134 irrigation facilities in the Mekong River Basin. Thailand ranks only second to Viet Nam by an irrigation area of 0.904 million ha in wet season, the share of irrigation area (in area of local district/county) is close to or larger than 10% in most regions. For spatial distribution, irrigation projects and irrigated regions are largely located in corridor zones along rivers and flood plains in northeastern Thailand in the Mekong River Basin. From the temporal perspective, wet-season paddy takes the largest share of irrigation area, nearly 95%; and the share of irrigation of dry-season paddy is the smallest one, less than 3%. Because of low runoff, high groundwater

Fig. 5.17 Huai Luang irrigation project in Thailand

level and soil salinization, the irrigation of dry-season paddy is restricted. Thailand is planning to build more irrigation projects in the northeastern region in the Mekong River Basin, with the aim to further advance the irrigation area and irrigated raio.

5.3.5 Viet Nam

Viet Nam has built up 670 dams/reservoirs for irrigation in the Highlands, 643 of which were identified as Earth dam. Viet Nam currently has 1360 irrigation projects in the Highlands, which could cover 47,001 ha; and 107 irrigation projects in the Delta region, which could provide benefit for 1.99 million ha. There are about 5625 earth canals with the total length of 13,410 and 23,110 km for the main and secondary canals, respectively, see Fig. 5.18. It indicates that nearly half the crop land in the Mekong Delta can be irrigated effectively. Similar with Cambodia, Viet Nam has three harvests of paddy a year—wet-season paddy, dry-season paddy and third-season paddy between them. At present, Viet Nam's irrigation projects mainly provide irrigation water to wet-season paddy and third-season paddy, and little irrigation water is provided to dry-season paddy and other crops. Viet Nam plans to build more small-scale irrigation projects in the eastern highland region. These irrigation projects will be used mainly to guarantee water supply for crops in non-paddy regions.

In Mekong Delta, saltwater intrusion increases the salinity of fresh water in the Mekong River, due to which the water can be hardly used for irrigation and "non-drought water deficit" is caused. Presently, Viet Nam has built a series of saltwater obstruction projects on small tributaries (e.g. Cai Oanh Tide Gate in Fig. 5.19), for the purpose of controlling the connection and exchange between inland rivers and saltwater and eliminating the influence of saltwater intrusion through structural approaches. Nevertheless, there is almost no tide gate on big tributaries, so the mainstream is still not exempted from the influence of saltwater intrusion. Besides, Viet Nam strengthens the coordination with upstream countries, and relieves drought influence through increasing water from the upstream.

5.3.6 Drought Relief in the Basin

The asymmetry of the upstream and downstream positions of countries along the Mekong River determines that these countries are not independent individuals, but are bonded by natural hydrological links. In order to meet the needs of drought control, upstream countries can usually transfer water from the Mekong River to tributaries for irrigation by constructing a large number of irrigation projects. This negative effect of "upstream mentality" giving priority to their own needs may lead to few water available in downstream countries, which in turn undermines their drought control capability. For example, in 2016, a large-scale meteorological and agricultural drought occurred in the Mekong River Basin. The Chinese government provided

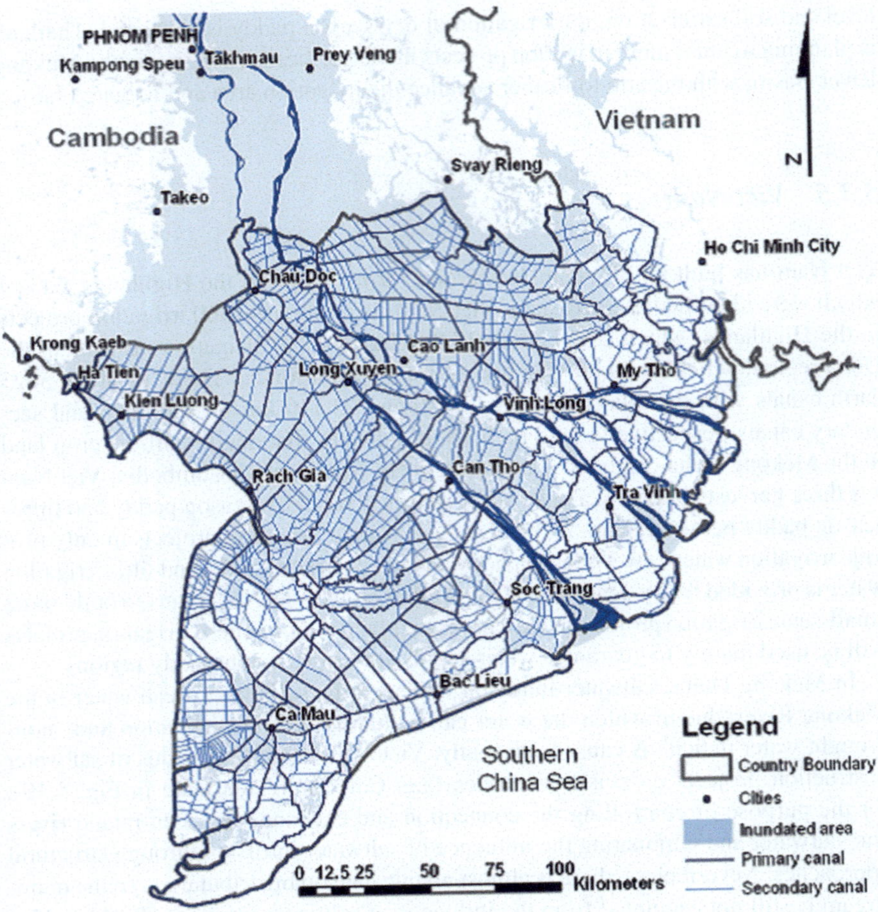

Fig. 5.18 Canal systems in the Mekong delta in Viet Nam (MRC 2005. Overview of the Hydrology of the Mekong Basin. Mekong River Commission, p 54)

emergency water supplement to the Mekong River by increasing the outflow of the Jinghong Reservoir. The research shows that during the emergency supply period, the net contribution of the discharge amount of Jinghong station to the total runoff of each hydrological station decreases from upstream to downstream. During this period, the runoff from upstream in Chiang Saen, Nongkhai and StungTreng stations respectively accounted for 99, 92 and 58% of the total runoff, and the net contribution rates of emergency water supply were 44%, 38% and 22% respectively. For the StungTreng station located downstream, the net contribution rate of emergency water supply is lower than the upstream stations, due to the water intake and use by countries in the middle and upper reaches, the consumption of water along the way and the inflow of tributary water. This indicates that only depending on the water supplement from upstream Jinghong station has a limited impact on drought relief downstream.

Fig. 5.19 Cai Oanh Tide Gate in the delta region of Viet Nam

Therefore, coordination and communication between countries along the Mekong River should be further strengthened, and scientific planning should be carried out from the whole basin level to further explore the greater potential for drought relief.

5.4 Non-structural Measures for Flood Prevention and Drought Mitigation

5.4.1 Mekong River Commission

As a cooperative organization for flood management of the Mekong River for countries in the basin, the Mekong River Commission (MRC) is responsible for collecting rainfall and hydrological data submitted by each country and forecasting floods of main rivers in the basin. In the flood season, between June and October each year, the Regional Flood and Drought Management Center (RFDMC) subordinate to MRC will release flood forecasts and alerts every day. Water levels of 22 predictive points in the Mekong River Basin are forecasted according to the data of 146 hydrological and meteorological stations of all countries in the basin. The RFDMC sends the daily bulletin to MRC, NOGs, media and the public. Figure 5.20 shows the water level forecast results of rivers in the basin for next 5 days. In Figs. 5.21, 5.22 and 5.23, Nongkhai River is taken as an example for an introduction to the measurement, forecast and comparison of its water level. Figures 5.24, 5.25, 5.26, 5.27, 5.28 and 5.29

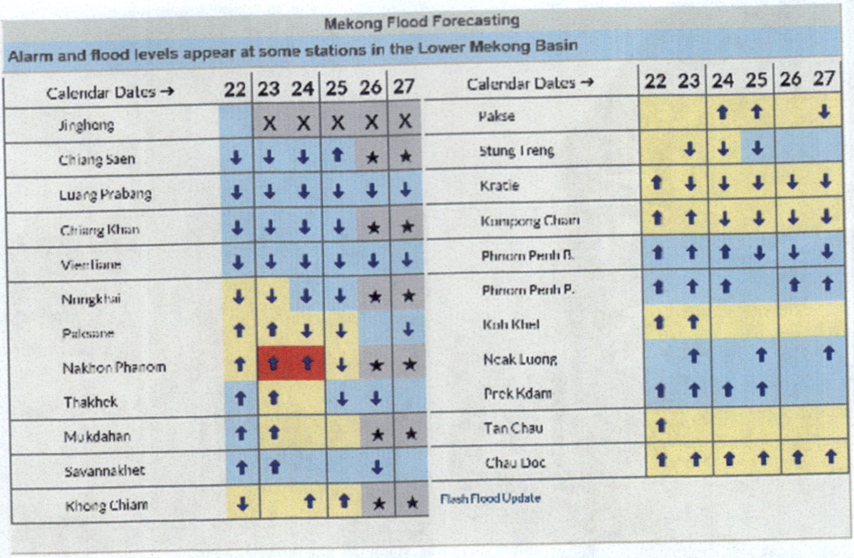

Fig. 5.20 Flood forecasting of MRC [http://www.mrcmekong.org (accessed on 18 July 2019)]

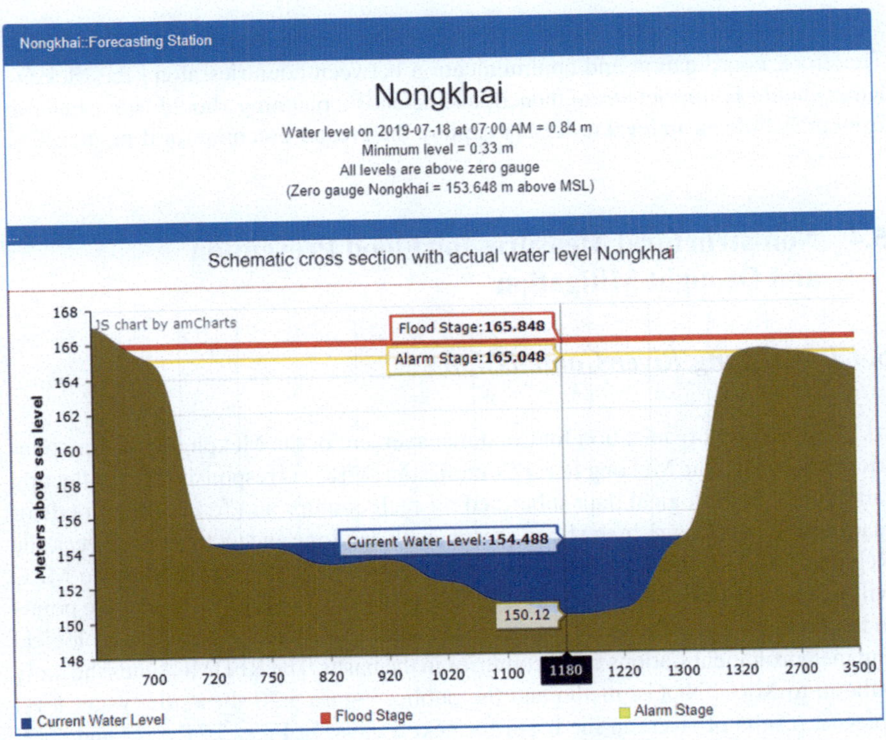

Fig. 5.21 One cross section and water level of Nongkhai River [http://ffw.mrcmekong.org/stations.php?StCode=NON&StName=Nongkhai (accessed on 18 July 2019)]

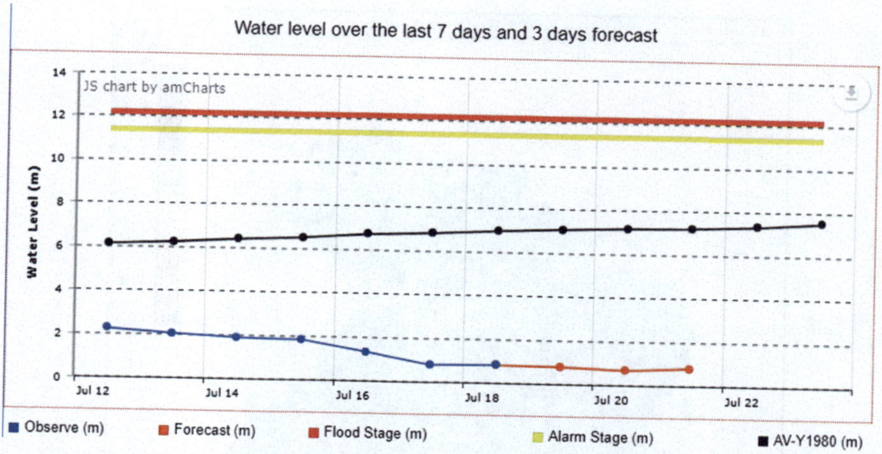

Fig. 5.22 Observed water level in the past 7 days and predicted water level in the next 3 days of Nongkhai River [http://ffw.mrcmekong.org/stations.php?StCode=NON&StName=Nongkhai (accessed on 18 July 2019)]

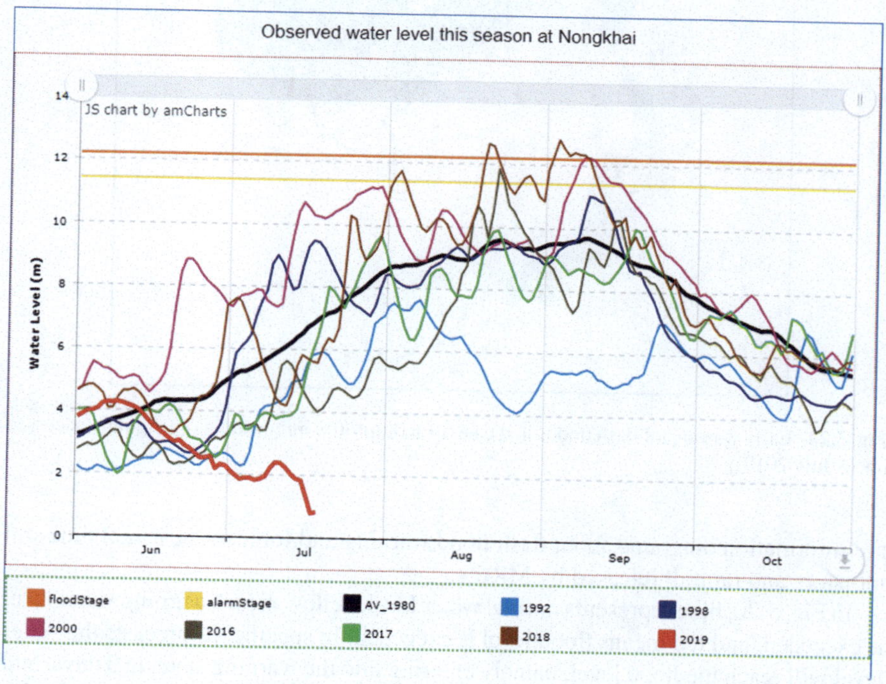

Fig. 5.23 Water level of Nongkhai station in flood season in different years [http://ffw.mrcmekong.org/stations.php?StCode=NON&StName=Nongkhai (accessed on 18 July 2019)]

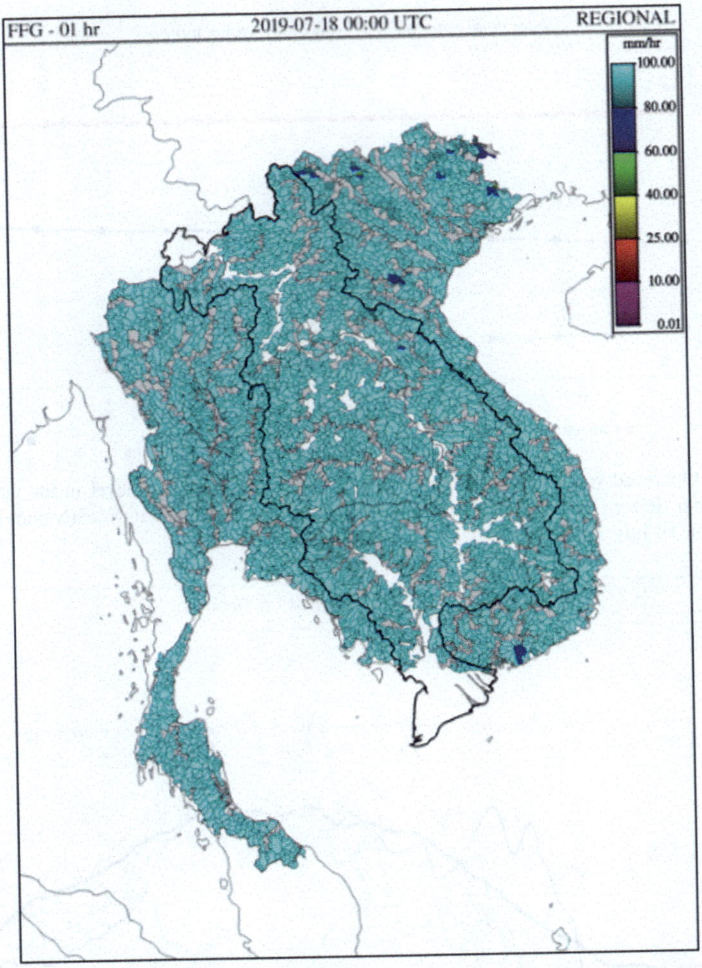

Fig. 5.24 Early warning of flash floods in the next 1 h [http://ffw.mrcmekong.org/ffg.php (accessed on 18 July 2019)]

are information diagrams about flash flood warning and forecast, near-real time soil moisture and rainfall released by MRC.

In Fig. 5.20, blue represents normal water level, yellow means warning water level is exceeded, and red means flood level is exceeded. In specific, it forecasts the water level will reach the flood level, namely entering into the warning state, in 3 days, and the flood level is determined by each member country.

In Fig. 5.23, blue line represents the measured water level, orange line represents the forecasted water level, red line is the flood level, yellow line is the warning level, and black line is the average annual level.

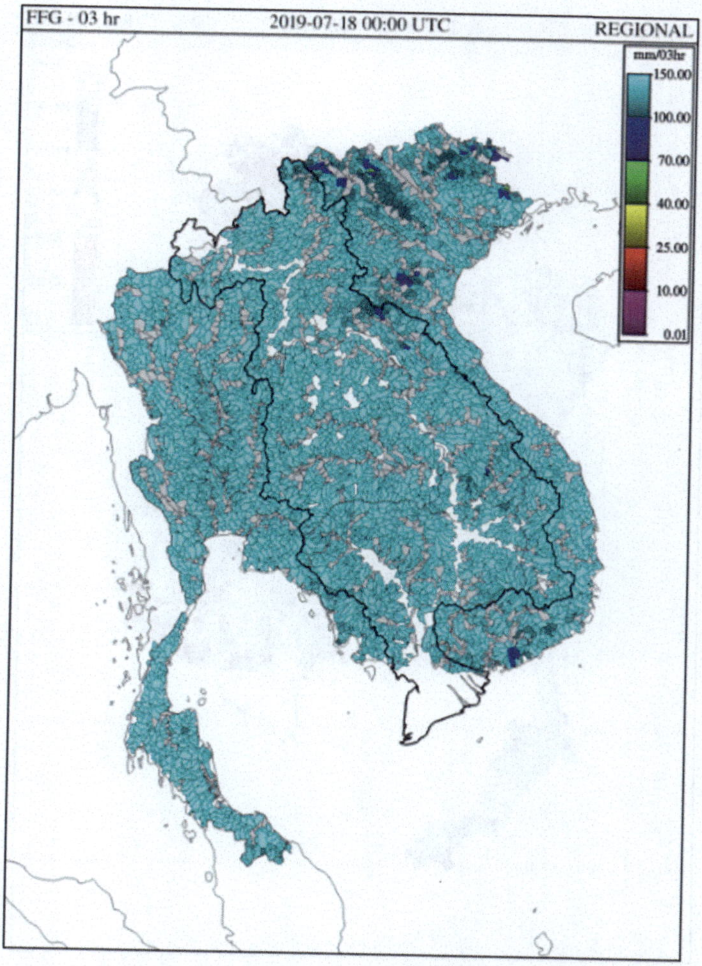

Fig. 5.25 Early warning of flash floods in the next 3 h [http://ffw.mrcmekong.org/ffg.php (accessed on 18 July 2019)]

5.4.2 Thailand

Thailand has a complete hydrological and meteorological monitoring, forecasting and warning system for the Mekong River Basin. There are 11 hydrological monitoring stations in the Mekong River Basin in Thailand. These hydrological stations are installed with the corresponding remote measuring system and video monitoring system, in which national hydrological monitoring technicians assure data can be sent to National Information System (NIS) accurately and in time. In the meantime, hydrological monitoring data can be transmitted in real time to MRC's server for data analysis, quality control and forecasting. Important stations forecast, for 3–5 days,

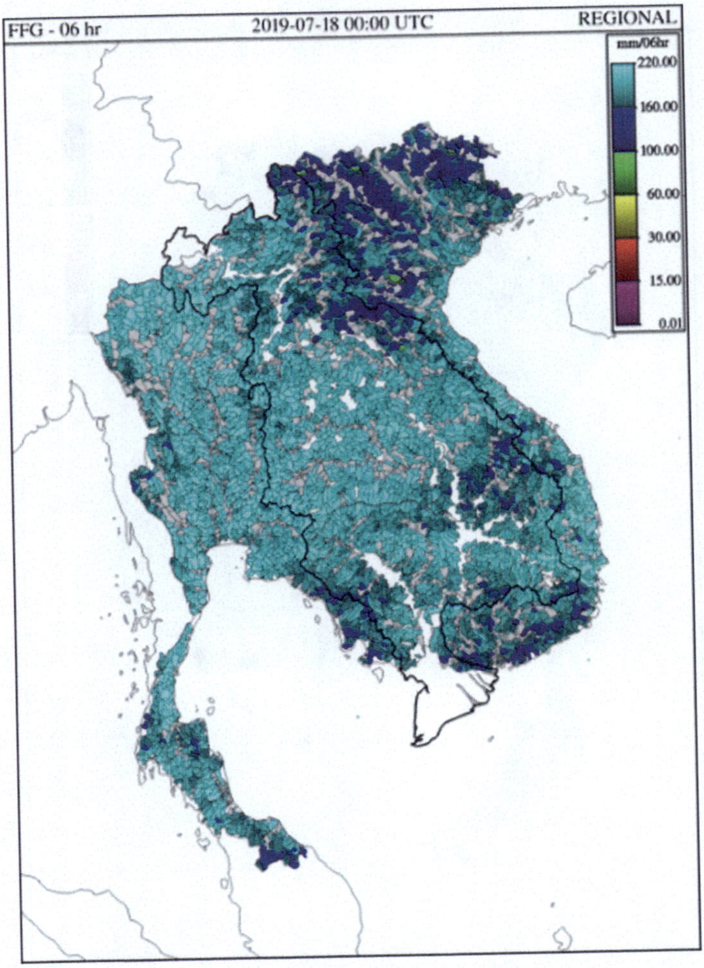

Fig. 5.26 Early warning of flash floods in the next 6 h [http://ffw.mrcmekong.org/ffg.php (accessed on 18 July 2019)]

water level and runoff every day and then, relevant information is released by NIS and MRC respectively to the general public and relevant departments. Meanwhile, the flash flood and rainstorm warning system is also established for mountainous regions, which can provide audible and visual alarm against rainfall, with green, yellow and red light alerts as well as tweeter against rainfall at different levels. After the alert is issued, villagers of the village where the alerting station is located will get evacuated to nearby highland and resettlement areas under the help of volunteers.

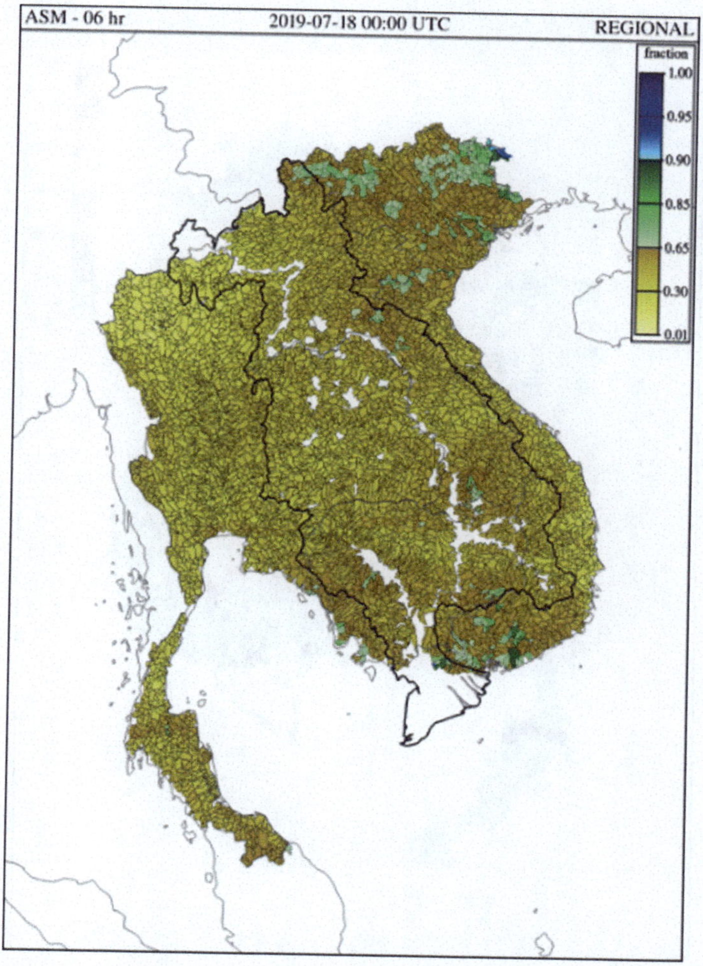

Fig. 5.27 Soil moisture in the past 6 h [http://ffw.mrcmekong.org/ffg.php (accessed on 18 July 2019)]

5.4.3 Viet Nam

The Mekong Delta plain in Viet Nam is low-lying without dyke or any other protective project. Even if there is any protective project, collapse banks are protected and reinforced, which are basically unable to withstand high flood level and are liable to high-level flooding of the Mekong River. In this region are some hydrological stations guarded by specially-assigned people, which can forecast automatically and send forecasting and warnings to MRC. National Mekong Committee of Viet Nam establishes a hydrodynamic model in the delta region that can forecast flood and send the forecasting to MRC for sharing. In the delta region, monitoring and forecasting of

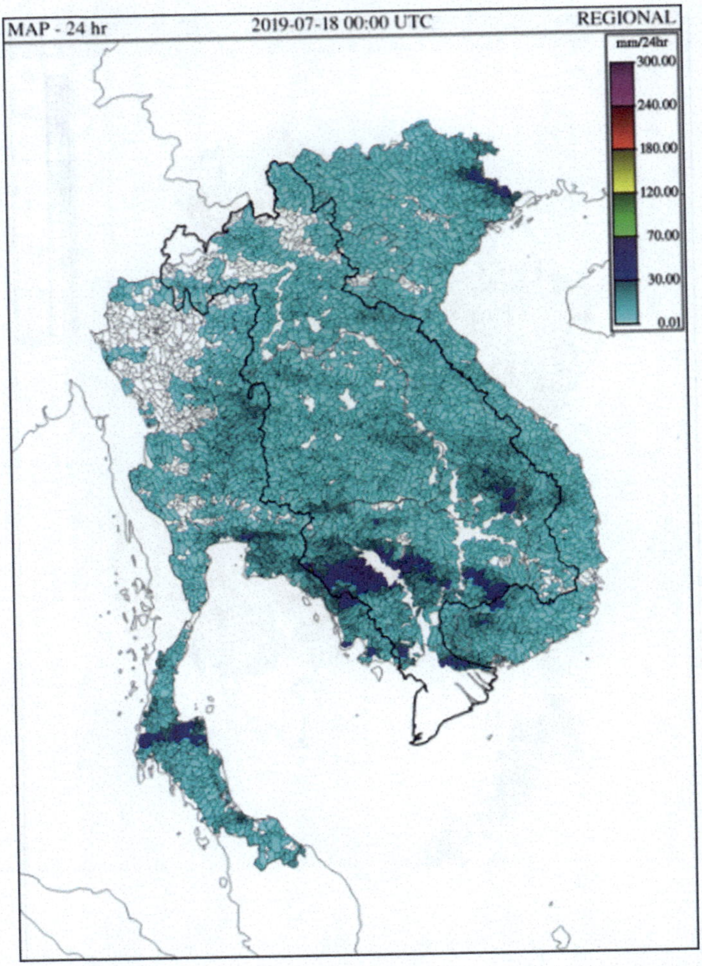

Fig. 5.28 Rainfall forecast in the next 24 h [http://ffw.mrcmekong.org/ffg.php (accessed on 18 July 2019)]

water regime is very important. When a flood takes place, monitoring and forecasting water level can early warn or transfer people in regions that may be flooded (Fig. 5.30).

In Viet Nam, flash flood becomes more frequent because of climate change in recent years. On average, 2–4 flash floods take place in the flood season every year. It usually occurs frequently in a place and features abruptness and great life and property losses in a small scope. At present, flash flood is hard to forecast, and prevention is carried out through designation of high risk regions and establishment of warning systems.

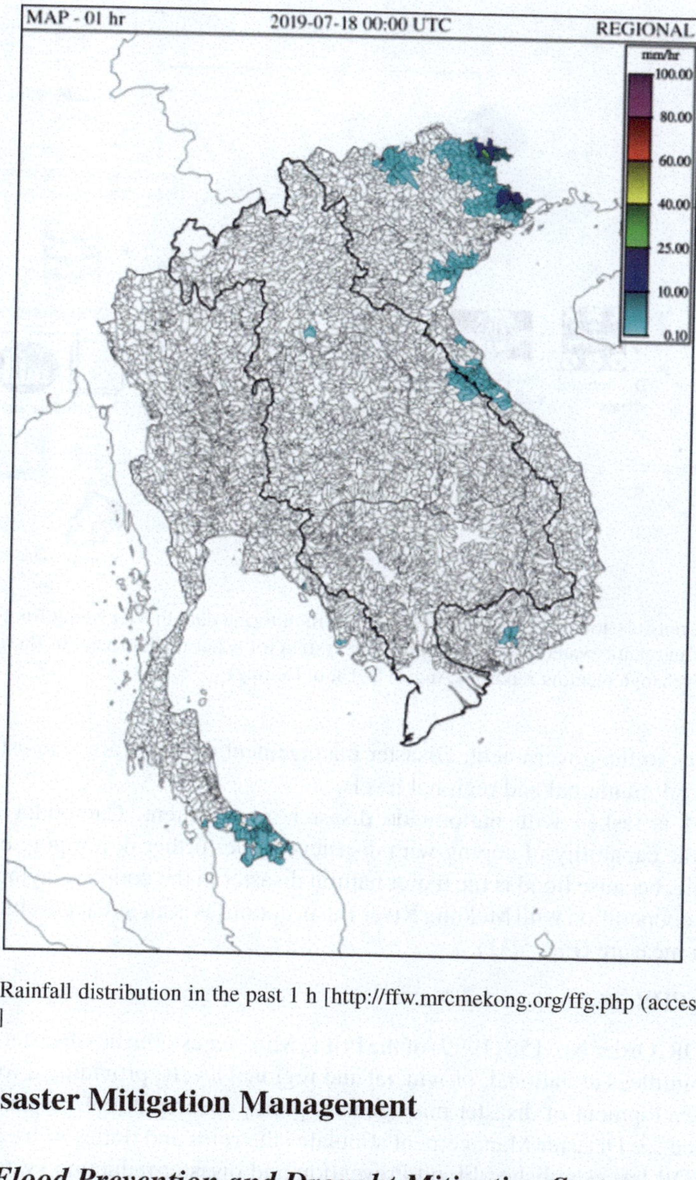

Fig. 5.29 Rainfall distribution in the past 1 h [http://ffw.mrcmekong.org/ffg.php (accessed on 18 July 2019)]

5.5 Disaster Mitigation Management

5.5.1 Flood Prevention and Drought Mitigation System

(1) Cambodia

Cambodia's disaster management body is the National Committee for Disaster Management (NCDM) founded in 1995. All departments and organizations should cooperate closely with NCDM and jointly provide disaster relief in emergency circumstances. As an executive unit, NCDM Secretariat provides disaster management

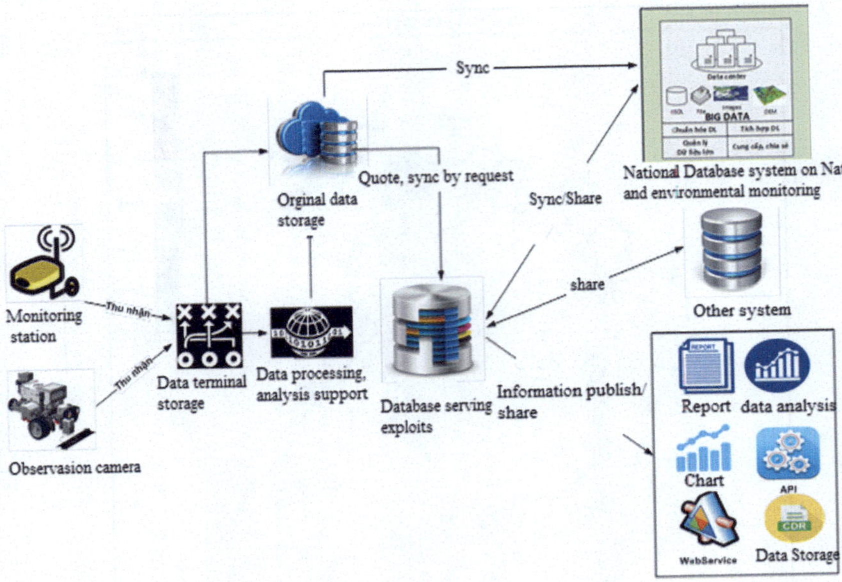

Fig. 5.30 Transmission and release of hydrological monitoring data in Viet Nam (Tong Thi Lien. Develop an automatic water resources monitoring system for water management in Ba river basin. Technical exchange meeting report in August 2018, in Beijing)

suggestions to the government. Disaster management councils are also established at provincial/municipal and regional levels.

NCDM is tasked with nationwide disaster management. Cambodia needs to improve the capability of coping with disasters while further developing economy. Meanwhile, because flood is the major natural disaster in the country, laying emphasis on the cooperation with Mekong River Basin countries is an indispensable disaster reduction measure (Fig. 5.31).

(2) **Lao PDR**

In Lao PDR, Order No. 158 (1999) of the Prime Minister establishes disaster management committees at national, provincial and regional levels, providing a foundation for the development of disaster management policy. Decree No. 97 of the National Committee for Disaster Management stipulates the roles and duties of its units.

Lao PDR has established Flood Prevention and disaster reduction systems from central government to towns and villages, with attention paid to the organization and coordination in flood prevention and mitigation. The coordination mechanism is in the charge of disaster management committees from national to basic levels. Although there is not specialized disaster management and mitigation law, duties of member units of the National Committee for Disaster Management are stated in relevant laws. In recent years, Lao PDR has started to attach importance to integrate city planning and waterlogging and implement uniform disaster relief measures (Fig. 5.32).

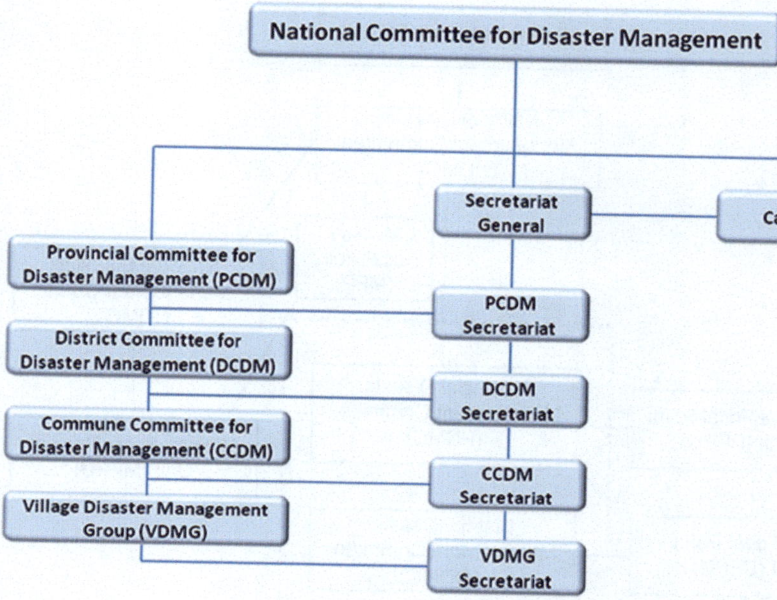

Fig. 5.31 Flood prevention and disaster mitigation system in Cambodia (https://www.adrc.asia/img_map/116_org.gif)

(3) Myanmar

In Myanmar, the major disaster management body is the Central Committee for Disaster Prevention and Relief (CCDPR). CCDPR was founded according to the guideline of the Safety and Management Committee of the National Peaceful Development Parliament, with the aim to implement disaster prevention and reduction measures more efficiently and led by the Minister of Internal Affairs and Religious Affairs. Subordinate to the Central Committee for Disaster Prevention and Relief are ten sub-committees on information and education, emergency communication, search and rescue, loss information and rescue, loss assessment and etc.

In the flood prevention and mitigation field, the Relief and Resettlement Department has cooperation with Department of Health, Department of Meteorology and Hydrology, Fire Services Department, Department of Human Settlement and Housing Development, Department of Irrigation and Myanmar Red Cross Society. The Department of Irrigation repairs and reinforces dams and watertight barriers in the delta region where flood often occurs, and builds cofferdams in several provinces liable to hurricane and storm surge. Once a disaster takes place, the Department of Meteorology and Hydrology will release to the general public information and alerts about hurricane, flood, super rainfall and river level.

In Myanmar, every stage of flood disasters is in the charge of specialized organization or department. In the flood prevention and disaster reduction field, the country

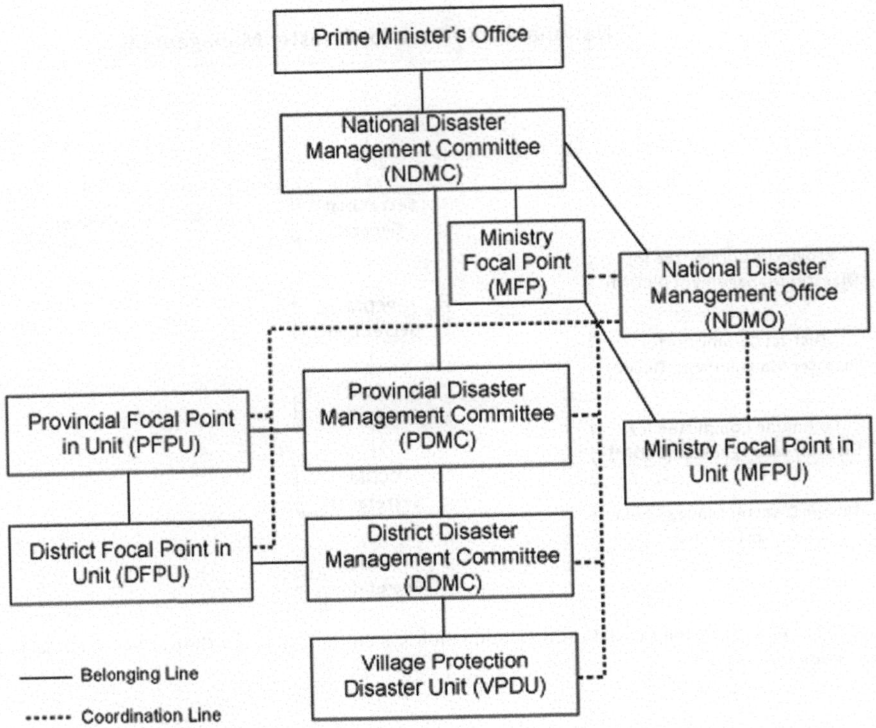

Fig. 5.32 Flood prevention and disaster mitigation system in Lao PDR (https://www.adrc.asia/img_map/418_org.gif)

attaches importance to the transmission of flood prevention information to the general public as well as the construction of structural facilities (Fig. 5.33).

(4) **Thailand**

Thailand promulgated the *Disaster Prevention and Mitigation Act* in 2007 and brought it into force on November 6 the same year. The country's flood disaster management system is founded on the basis of this act. It has 5 distinguishing features: 3 major decision-making and planning bodies are established at national level, provincial level and of Bangkok City; Prime Minister or appointed Vice Prime Minister serves as Commander at national level; the Department of Disaster Prevention and Mitigation (DDPM) is authorized to be the core department for flood disaster management; local governments are authorized to be responsible for local flood disaster management according to provincial-level planning (Fig. 5.34).

Presently, Thailand's national overall plan for disaster prevention and mitigation is divided into three parts. Part I includes flood situation and management system, with description of the situation of flood disasters, management system, planning objectives and emergency work for coping with flood disasters. Part II sets forth

Fig. 5.33 Flood prevention and disaster mitigation system in Myanmar (https://www.adrc.asia/img_map/104_org.gif)

standard operation procedures. Part III discusses national safety issues related to flood prevention and mitigation.

The Thailand National Committee on Disaster Prevention and Mitigation (NCDPM) is a major department responsible for formulating policy proposals, mainly including developing national plans for disaster prevention and mitigation and arranging the disaster prevention and mitigation abilities of central government, local governments and other related private sectors. NCDPM is led by Prime Minister and consists of 34 members from departments, institutions and organizations related to disaster management. At national level, Prime Minister or appointed Vice Prime Minister serves as President of NCDPM, responsible for formulating disaster risk management policies. At local level, provincial committees on disaster prevention and mitigation are established.

(5) **Viet Nam**

At present, Viet Nam is striving to probe into the legislation of flood disaster management and gradually establish a set of legal system. Relevant legal documents formulated and issued in recent years include the Dyke Management Law, Law on Water Resources, Forest Protection and Development Law, Environmental Protection Law, Land Law, Ordinance on Dyke Management, Ordinance on Flood and Storm Control, Ordinance on Development and Protection of Water Resources Structure and Ordinance on Structure of Hydrometeorological Survey.

The Central Committee for Storm and Flood Prevention (CCSFC) is a major institution that formulates flood management regulations and mitigation measures

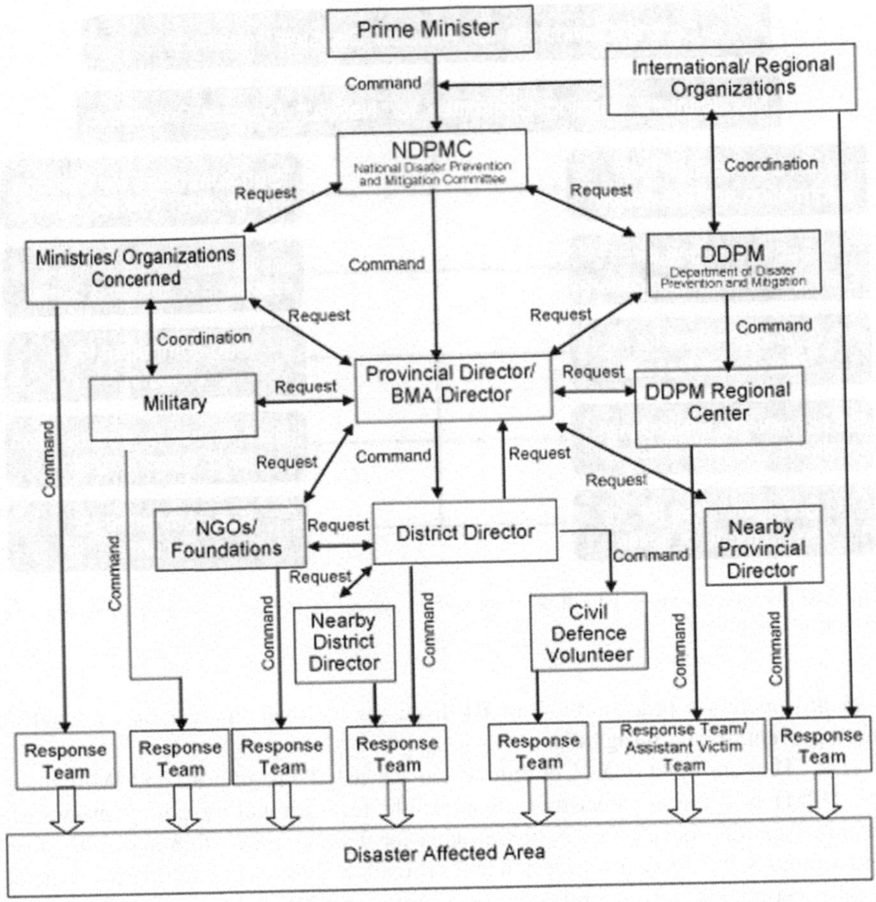

Fig. 5.34 Flood prevention and disaster mitigation system in Thailand (https://www.adrc.asia/img_map/764_org.gif)

of Viet Nam. At the central level, CCSFC is responsible for coordination of disaster relief. CCSFC Secretary is appointed by the Dyke Management and Flood Prevention Department of the Ministry of Agriculture and Rural Development. CCSFC is responsible for formulating flood regulations and mitigation measures, and local emergency work is coordinated by storm and flood mitigation committees at provincial level. As the highest coordinating body, CCSFC is tasked with nationwide flood prevention and disaster reduction. It is the primary unit that implements relevant policies.

For CCSFC, the most import task is the upgrading and maintenance of the dam system including river dams as long as 5000 km and seaside dams as long as 3000 km. Under the help of the United Nations Development Programme (UNDP) and the World Food Programme (WFP), Viet Nam is restoring and reinforcing the dykes on northern and central coasts. In addition, Viet Namese authorities have set about

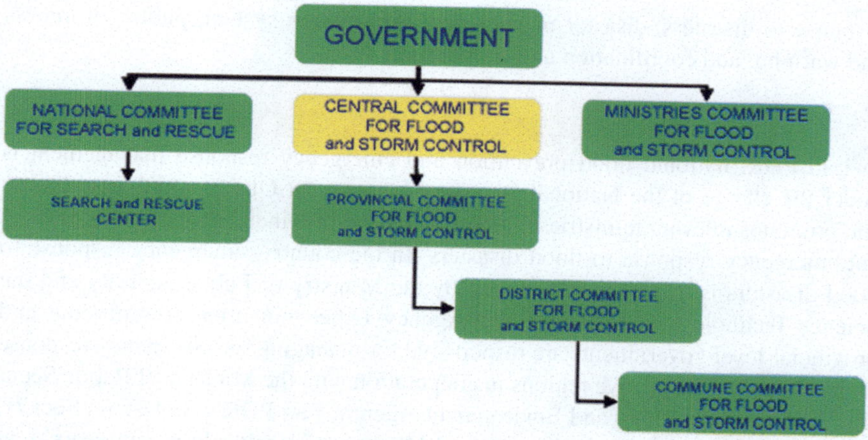

Fig. 5.35 Flood prevention and disaster mitigation system in Viet Nam (https://www.adrc.asia/img_map/704_org.gif)

reinforcing dykes in the Red River Delta Region, and have started to prepare and improve flood discharge plans of important reaches (Fig. 5.35).

Along with national economic development, Viet Nam is improving its flood disaster coping mechanism in dykes, forest protection and development, water resources utilization and other links. In the meantime, Viet Nam also lays emphasis on improving communities' risk prevention awareness. It develops and implements a string of social and economic development projects with the focus on flood management, such as water conservation forests and protective forest belts. Besides, more attention is also paid to the all-round management of flood and drought in such projects as reservoir construction for coping with flood and drought, the project of "Living with Flood" and reinforcement of dykes.

5.5.2 *Emergency Response to Floods and Droughts*

(1) **Cambodia**

Cambodia recognizes it's urgent to develop integrated disaster management strategies after going through the recent destructive flood and realizing similar events may continue to take place in future. At present, Cambodia's comprehensive disaster management strategies include "flood recovery plan" and the local project of rural disaster management "community independence and flood mitigation plan". The flood recovery plan aims to recover the socio-economic structure and in the meantime, to indirectly support production and income recovery in rural areas. The "community independence and flood mitigation plan" is a technological assistance project that helps the Cambodian government improve community independence during periodic floods. Four tasks are included: preparedness for and emergency response to disasters, disaster management information system, public awareness and warning, and coordination and cooperation.

(2) **Lao PDR**

In Lao PDR, national flood prevention and emergency response management is under the charge of the National Disaster Management Office, which coordinates and promotes relevant ministries, commissions and provincial governments to carry out emergency response to flood disasters. In the country, emergency response to flood disasters is usually implemented by the Ministry of Public Security and the Science Technology and Environment Agency. Other ministries, commissions and provincial-level governments are responsible for managing the emergency response to floods in their respective regions in cooperation with the Ministry of Public Security, Science Technology and Environment Agency, Lao PDR's Red Cross Society, Ministry of Agriculture and Forestry and Ministry of Health. In recent years, Lao PDR has started to lay emphasis on integrating city planning and construction with urban flood disasters and implementing uniform measures of emergency response to flood disasters.

(3) **Myanmar**

Fully recognizing the importance of international cooperation in flood prevention and mitigation, training and experience exchange, the Myanmar government often sends officials to participate in training courses, study classes and relevant meetings in foreign countries. On the other hand, it holds study classes and workshops through cooperation with or sponsorship of international and regional organizations such as the UN Economic and Social Commission for Asia and the Pacific (ESCAP), World Meteorological Organization (WMO) and UN Development Programme (UNDP).

(4) **Thailand**

Thailand Department of Disaster Prevention and Mitigation (DDPM) is a major disaster management institution founded in 2002 and subordinate to Ministry of Interior. It is responsible for the coordination of departments related to disaster prevention and mitigation. According to law, Minister of Interior should be Commander in Chief in emergency circumstances of disasters, especially large-scale disasters. DDPM Director General is NCDPM Secretary General. When a disaster takes place, relevant officials should be on-site commander.

Related decision-making organizations, at national level, provincial level and of Bangkok, are led by Prime Minister or appointed Deputy Prime Minister, provincial governor and Mayor of Bangkok respectively. Decision-making organization at each level includes the corresponding disaster prevention and mitigation committee. Director at provincial level can ask for support from neighboring provinces according to disaster condition. Regional director can ask emergency teams, civil defense volunteers and even NGOs to participate in disaster relief. Local emergency units, such as civil defense centers, should respond rapidly to disasters; disasters with which emergency units are unable to cope should be dealt with by competent units, such as civil defense centers and police and fire stations; backup teams should be responsible for manpower, equipment and technology support according to the requirements of emergency units or competent units, and all government organs, NGOs, regional civil defense centers and armed forces are backup units.

Subordinate to DDPM, the Disaster Prevention and Mitigation Academy (DPMA) carries out flood disaster management trainings oriented to practitioners, managers and local government officials. DPMA is founded by the Ministry of Interior. It has become the most important flood management education base, and carries out flood prevention and mitigation education actively in elementary schools.

(5) **Viet Nam**

Viet Nam attaches great importance to public involvement and use of resources of international organizations. When a disaster takes place, government, labor unions, youth and women associations and other social organizations will organize donation activities proactively to help victims in affected areas. In disaster relief and recovery, emphasis is laid on mutual assistance. Public awareness is improved, especially through media, so that officials at basic level and from various departments and local governments receive a good training. As a result, awareness of government officials and the general public is improved.

The government provides preferences to flood disaster projects every year, and continues to increase disaster mitigation spending. Priority is given to flood prevention and mitigation plans and programs, such as afforestation project, dyke renovation project, reservoir project, landslide project, "Living with Flood" project and etc.

5.6 Assessment of Flood Prevention and Drought Relief Capacity

5.6.1 Cambodia

Because flood is a major disaster in the country, Cambodia has a relatively complete organizational system of flood mitigation, with disaster management departments at national, provincial and community levels. Non-structural measures for emergency response to flood are implemented through some "flood management programs" and "disaster mitigation programs". Limited to economic and other factors, however, apart from part of the revetment works in cities such as Phnom Penh, Kratie and Kampong Cham, there is no flood prevention facility established actively in most regions. Though 14 reservoirs are planned with a gross storage capacity of 18.89 km^3, only one has been buit up, with a storage capacity of 120,000 m^3. Based on the population and planned storage capacity of Cambodia within Mekong River Basin, the per capita reservoir storage capacity will be 1629 m^3/people, a little higher than the world average level of 1000 m^3/people.[2] But the current per capita reservoir storage capacity is only 0.01 m^3/people, much lower that of the world average. Residents surrounding the Tonle Sap Lake and flood plains of Mekong River live on fishery and catch and are accustomed to flood. On the contrary, flood reduction is regarded as damage to their life. Rainfall is small and flash flood is less destructive in the mountainous regions in upper Tonle Sap Lake in western Cambodia, but eastern mountainous regions are liable to floods from upper mountainous regions in Lao PDR and Viet Nam. In the meantime, the provision of huge capacity of flood storage and detention in the Tonle Sap Lake and flood plains (Tonle Sap lake of 70 km^3, floodplains of 35 km^3) of Mekong River reduces the flood prevention pressure of the delta region in the lower reaches in wet season and also provides continuous water replenishment in dry season. In recent years, along with the improvement of flood prevention capacity in upper regions and the increase of irrigation water diversion, flood level of the Tonle Sap Lake and flood plains of Mekong River in Cambodia has been no longer as high as in history (e.g. 2000), so some regions in the flood plains of Mekong River, especially near Phnom Penh, have been developed, such as reclamation and banking for infrastructure construction, which reduces the flood storage capacity of the flooded area to some extent. But the development should be a natural balance process between water and human as long as it is conducted within a scope.

For extremely uneven spatial distribution of irrigation engineering system, the drought relief capacity of structural measures distinguished much among regions within Cambodia. Presently, irrigation coverage rate exceeds 10% universally in the region centering on capital Phnom Penh and Tonle Sap Lake, enabling the region with strong drought relief capacity. However, there is no irrigation project in northeastern, northwestern regions with no irrigation area. Though one reservoir of 120,000 m^3 storage capacity has been buit, the irrigation function of it is very limited. From the perspective of financial support, the per capita GDP of Cambodia is 1270 USD, ranking 158 in the world, much less than the world average. Comparatively limited financial support constrains the government input to drought relief. Based on the above two aspects, the drougt relief capacity of Cambodia is not outstanding, and there are great differences between different regions.

5.6.2 *Lao PDR*

Lao PDR has established a flood prevention and mitigation system from central government to towns and villages. Lao PDR National Disaster Management Office is responsible for organizing and coordinating emergency management of disasters and enabling emergency response to flood. In Lao PDR, the flood prevention engineering system is weak on the whole, and banks along the Mekong River, except some big towns and cities with revetment, are basically in a natural state. The revetment

[2]Hydropower and Dams, World Atlas, 2008.

account for less than 1% of the whole bank. Therefore, riparian regions are easily flooded once water level of Mekong River is high. Except for cities like Vientiane, Pakse, Savannakhet, Champasak are installed with sluices and pump stations for flood prevention, tributaries of Mekong River in Lao PDR are merely installed with flood prevention project, so backflow effect occurs easily in case of high water level in Mekong River and floodwater may affect large areas in upper tributaries. In most mountainous regions, rainfall runoff is frequently and flash flood takes place very easily, but the flash flood warning construction gradually conducted in recent years can relieve the damage of flash flood to some extent. The existing, under construction and planned reservoirs in Lao PDR have a gross storage capacity of 58.62 km^3. According to the population and planned (including the existing and under construction) reservoirs of Lao PDR within Mekong River basin, the per capita storage capacity is 11,060 m^3/people, much higher than the world average (1000 m^3/people), after Norway who ranking number two in the world (14,000 m^3/people, see Fig. 5.36); the per capita storage capacity based on the existing and under construction projects is 5507 m^3/people, which is 5.5 times of the world average, and close to than of German (see Fig. 5.36). Therefore, construction of these reservoirs could control some floods. However, these reservoirs are not designed for flood prevention, so floodwater can be only discharged instead of being stored. After the collapse of the Xe Pian-Xe Namnoy Reservoir in 2018, adjustment of policy may help to promote the flood prevention function of reservoirs.

Per capita GDP of Lao PDR in 2016 is 2339 USD, ranking number 138 in the world, much less than the world average. With limited investment to water conservancy projects in history, Lao PDR has only 2000 plus irrigation projects in total,

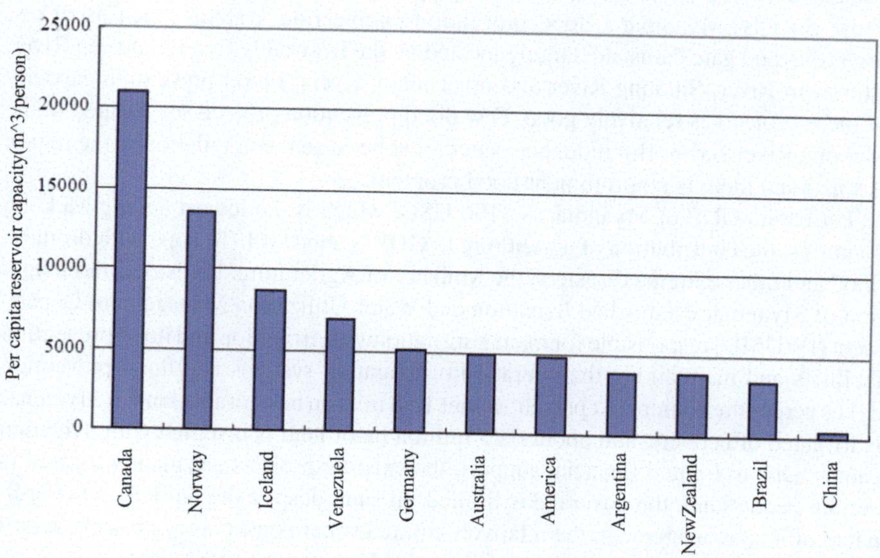

Fig. 5.36 Per capita reservoir storage capacity of typical countries in the world (http://www.chincold.org.cn/chincold/upload/news/lin20091125453454 2.pdf)

due to which the irrigation percentage is less than 1% in nearly 95% territorial area of Lao PDR. Meanwhile, most of these irrigation projects are of a small size, less than 200 ha, and located in riparian area along tributaries and in small floodplains of lower reaches of Mekong River. At present, 10 reservoirs have been built on Mekong river tributaries in Lao PDR, but they are used mainly for power generation, and the contribution to irrigation is negligible. Overall, the irrigation projects are with limited quality and uneven distribution, as a result, the water supply per unit area of cultivated land and the irrigation percentage are low, which affects the water conservancy guarantee ability of drought resistance. It should be noted that Lao PDR is exerting great efforts to invest in water conservancy projects, and will build another 2768 irrigation projects and increase irrigation area by 240% for non-paddy fields and 460% for dry-season paddy by 2030 and thereby, improve the drought resistance capacity.

5.6.3 Myanmar

In Myanmar, the Central Committee for Disaster Prevention and Relief is a major body of disaster prevention and mitigation that has multiple sub-committees of information and education, emergency communication, search, loss information and rescue, loss assessment and etc., and has cooperation in emergency rescue and response for disaster prevention and mitigation with Department of Health, Department of Meteorology and Hydrology, Fire Services Department, Department of Human Settlement and Housing Development, Department of Irrigation and Myanmar Red Cross Society. Myanmar's flood prevention engineering systems, such as dykes, reservoirs and gate dams, are largely located on the Irrawaddy River, Salween River, Chindwin River, Sittaung River and other major rivers. Flood prevention capacity of these projects is relatively good. Few flood prevention projects are located in the Mekong River Basin. But influence is not great because the population in the region is small and there is no prominent flood problem.

Per capita GDP of Myanmar is 1196 USD, which is the lowest among Mekong countries, the contribution of agriculture to GDP is about 1/4. To cope with drought, flood and other extreme disasters, the Ministry of Agriculture, Livestock and Irrigation of Myanmar established Irrigation and Water Utilization Management Department (IWUMD) responsible for managing nationwide irrigation and flood prevention facilities and maintaining the operation of irrigation systems and flood prevention dykes across the country. At present, about 1.58 million ha of arable land in Myanmar is irrigated effectively, and about 0.25 million ha of land is installed with irrigation canals. Due to limited financial support, the existing projects are mainly located in certain paddy land, the coverage is limited. In sum, despite the unified and coordinated official management, the relatively limited water conservancy project support capacity has limited the drought resistance of Myanmar to some extent.

5.6.4 Thailand

Thailand has a complete meteorological and water level monitoring and warning system that can monitor rainfall and water regime in real time, provide relevant forecasting, and warning against flash flood. Its disaster mitigation and relief system is also sound, covering organizational systems of emergency response at national, provincial, regional and response team levels. Most tributaries (at least 5) on Korat plateau are installed with sluices and dams, having regulation capacity of tributary flood and Mekong flood. The left tributaries without sluices are apt to influence of high water level of Mekong River. Urban areas along Mekong River are mostly protected by revetments, 300 km out of the total length (955 km) of the Mekong mainstream in Thailand are revetment or embankment works, which could withstand the high water level of Mekong River. But the natural low banks are still apt to flooding during the high water level period of Mekong River. Thailand has not set up a specific agency to in charge of the national flood defence coordination, resulting in weak national coordination ability. The gross storage capacity of the existing and under construction reservoir in Thailand is 3.57 km^3. The per capita reservoir storage capacity is 155 m^3/people, lower than the world average of 1000 m^3/people. For flood prevention capacity is not designed for these reservoirs, they could not play an important role in flood detention and regulation.

Per capita GDP of Thailand in 2016 is 5980 USD, ranking the top among Mekong countries. With strong financial support, Thailand has established a full-coverage and relatively complete hydrological and meteorological monitoring and forecasting system, including 11 major hydrological stations on mainstream Mekong River, which is used to get real-time water (e.g. water level, runoff, etc.) information, diagnoses occurrence of flood and drought in time, and provide scientific decision-making support for emergency dispatching of irrigation water. With regard to engineering support, Thailand ranks top in countries along Mekong River by the quantity of irrigation projects, thanks to which nearly 14% of northeastern Thailand (in Mekong River Basin) is covered with reservoirs, weirs, channels and other water conservancy facilities. So dense irrigation projects assure the share of irrigation area in northeastern Thailand exceeds 10%, and provide reliable emergency water sources for local crops' production. On the whole, Thailand is at a high level and has a good capability of drought resistance because of good capacity in drought prevention and mitigation, powerful economic support capacity and water resources guarantee capacity. Nevertheless, it should be pointed out that current irrigation aims mainly at wet-season paddy, and the irrigation guarantee capacity for dry-season paddy remains insufficient. As a matter of fact, drought is the commonest natural disaster in the dry season in Thailand that seriously influences the growth of dry-season paddy and other crops. Therefore, improving the irrigation guarantee capacity in dry season is critical for improving drought resistance level and capacity of Thailand.

5.6.5 Viet Nam

Viet Nam has a sound legal system for disaster prevention and mitigation. Disaster mitigation committees at national, provincial, regional and community levels were established. And the capability of emergency response to disasters is improved through improving public awareness and personnel trainings and drills. Mekong Delta is a major food production region and aquaculture region of Viet Nam that is liable to the influence of flood, drought and other disasters. With a low-lying terrain and few dykes and other flood prevention projects, the region is definitely liable to long-term and large-scope flooding when water level of Mekong River is high. In recent year, Viet Nam has avoided flooding through migration or lifting residential housing foundation, which is also an active preventive measure. In addition, if reservoirs on upper Mekong River are all capable of flood prevention, flooding of the delta region may be relieved to some extent when regulation is conducted well and floodwater is stored normally in Tonle Sap lake and floodplains along Mekong River in Cambodia. There are more than 10 reservoirs on Se San River and Ea Krong River in Central Highlands on tributaries of Mekong River, with a gross storage capacity of 3.16 km^3. The per capita reservoirs storage capacity of Viet Nam within the Mekong River basin is 152 m^3/people, lower than the world average of 1000 m^3/people. Nevertheless, these reservoirs are desined mainly for power generation without flood prevention capacity, could not play a role of flood storage and regulation.

Mekong Delta is "a land flowing with milk and honey" in Viet Nam, where drought is a major natural disaster. Per capita GDP of Viet Nam in 2016 is 3179 USD, after Thainland. Besides, the Viet Namese government establishes a uniformly coordinated drought resistance and mitigation system, which assures the scientific formulation and implementation of flood prevention measures. In the region, because of the construction of a set of large irrigation facilities, Viet Nam ranks top in countries along Mekong River by irrigation area (nearly 2 million ha or 50%). It means nearly half the paddy fields in the region can receive water replenishment through irrigation. This region is faced with typical drought caused by the shortage of freshwater because saltwater intrusion leads to soil salinization and water salinity rises which cannot be used for irrigation. Hence, Viet Nam prevents saltwater from flowing into inland rivers through building tide gates and thereby, withstands freshwater shortage-caused drought. Overall, Viet Nam has good resistance ability against drought to make up the locational weakness of Mekong Delta. It is notable that due to limited financial input, the existing tide gates are mainly located on small tributaries, making the mainstream and riparian area apt to salinity intrusion. Better resolvement of salinity intrusion problem is critical for further improving the drought resilience of Viet Nam.

5.6.6 Overall Situation at the Basin Level

In general, the overall flood prevention capacity of the Mekong River Basin is weak. The Tonle Sap Lake and the Mekong River floodplain in Cambodia provide large flood prevention capacity. Due to the insufficient flood prevention engineering system, it is inevitable that it will be exposed to flood damage. Lao PDR is mostly hilly and mountainous. Due to the weak flood prevention works along Mekong River, it is more vulnerable to floods from the mainstream. Myanmar is in the upper mountainous area in the Mekong River Basin. Although there is no flood prevention engineering system such as levees on mainstream Mekong River, the flood hazard is relatively small due to the scarcity of riparian population. Despite the high terrain, and some flood prevention projects such as revetment, sluice and pump, in the case of regional heavy rain and the high water level of Mekong River, it will still suffer serious damage. Viet Nam is at a disadvantage in flood prevention since it is in the delta area. In addition, its flood prevention works such as embankment are weak. So it is easy to suffer from serious flooding when floods from the upper reaches of Mekong River encounter the high tides.

The situation in the Mekong River Basin countries varies greatly. They are afraid of the flood of the Mekong River, while but they need the water from it. Without an effective coordination and cooperation mechanism, flood prevention at the basin level is hard to be effective. The flood management cooperation mechanism adopted by some transboundary rivers in the world may be worth learning, for example the Rhine River in Europe. It originates from the northern foothills of the Alps in Switzerland, flows through six countries including Switzerland, Liechtenstein, Austria, France, Germany and the Netherlands, and goes into the North Sea near Rotterdam in the Netherlands, with a total length of about 1232 km. At the international level, an institutionalized cooperation mechanism has been established for the Rhine River and ambitious flood prevention targets set (Rhine Action Plan: ICPR, 1998) to provide framework for joint flood management in the whole basin. Flood risk management measures for the whole basin include flood risk zoning, coordination mechanisms between countries, and economic compensation schemes between upstream and downstream regions. The results of many years of operation show that the cooperation mechanism and framework of the Rhine River, a transboundary river, has been successful.

The Lancang-Mekong River flows through China, Myanmar, Laos, Thailand, Cambodia and Viet Nam and goes into the South China Sea in the Delta of Viet Nam. The total length of the Lancang-Mekong River is about 4880 km, which is much longer than that of the Rhine River. Since it flows through many countries, it also faces the problem of flood prevention cooperation and coordination among countries. Therefore, the successful and advanced flood management experience of similar transnational rivers such as the Rhine could be learned from and establish a flood prevention cooperation mechanism and framework at the basin level. The flood prevention structural facilities in the basin could be strengthened, other activities like improving the flood prevention capacity, carrying out flood risk zoning and assessment, developing flood detention and control functions of reservoirs in the basin, coordinating flood regulation between upstream and downstream countries to make full use of flood resources, giving full play to their respective regional and geographical advantages, and appropriately adopting economic adjustment levers could be taken to achieve unified flood management of the whole basin.

5.6.7 *Joint Efforts to Cope with Flood and Drought*

For all regions of the world, water security is a prerequisite for economic and political stability in any country and region. In order to achieve water security, countries in the basin must establish mutual trust and cooperation, have a unified understanding of each other's rights and obligations, and build a platform for efficient cooperation in science and technology, information sharing and basin management.[3] The Lancang-Mekong River Basin countries are exploring the establishment of a whole basin cooperation mechanism and have carried out a lot of cooperation in the field of disaster prevention and relief.

In 1995, Laos, Thailand, Cambodia and Viet Nam jointly established the Mekong River Commission (MRC). The MRC and its member countries have played an important role in the sustainable use of water resources in the Mekong River Basin, especially the Regional Flood and Drought Management Center (RFDMC) for forecasting and warning the floods of the Mekong River, which is of great significance for flood prevention in the basin. China and Myanmar have maintained close cooperation with the four downstream countries through dialogue mechanism of the MRC. Since 2002, China has regularly submitted to the MRC the hydrological data of the Lancang River in flood season and the emergency hydrological information during the dry season. In 2010, in order to help downstream countries cope with the extreme drought, the Ministry of Water Resources of China provided hydrological information to the MRC for emergency response. Later, when encountering various special climates such as typhoons, the Chinese side repeatedly informed the MRC the regulation information of the upstream reservoir. The MRC and its member countries highly valued the support from China and considered the hydrological information provided by China is of great significance to the flood prevention and drought relief work of the lower reaches of the Mekong River.

Information sharing and technical cooperation have also been carried out among the countries in the Lancang-Mekong River basin. For example, China has signed memorandum of understanding on water resources cooperation or agreement on reporting flood information with Laos (2014), Thailand (2014 and 2016), Cambodia (2013), and Viet Nam (2002). In recent years, the Chinese side invites more than 100 water conservancy officials and experts from downstream countries and MRC to visit China every year, to visit the reservoirs in the Lancang River, the Three

[3]Zhong et al. Rivers and reciprocity: perceptions and policy on international watercourses, Available Online 29 February 2016, wp2016229; 10.2166/wp.2016.229.

Gorges Project, the South-to-North Water Diversion Project, and the water conservancy projects in Beijing, Shanghai, Guangzhou, and to share with them Chinese experience. At the same time, the Chinese side has also sent several delegations to the Mekong River Basin to carry out technical exchanges and understand the concerns of downstream countries.

In the case of major disasters, the countries in the Lancang-Mekong basin actively carried out disaster alleviation cooperation and jointly cope with the challenges of floods and droughts. For example, the Chinese government implemented emergency water supplement in response to the major drought in the Lancang-Mekong River Basin in 2016 caused by the El Nino phenomenon. According to the joint assessment of emergency water supplement effect carried out by Chinese side and Mekong River Commission, the emergency water supplement from the Lancang River increased the amount of water flowing into the Mekong River by about 1000 m^3/s, which raised the water level along the main stream of Mekong River by 0.18–1.53 m. A letter from the Mekong River Commission Secretariat CEO Dr. Pham Tuan Phan to the Minister of Water Resources of China stated: "*the Secretariat views this decision (emergency water supply) as a kind and considerate gesture exhibited by a good neighbor and friend to the Mekong countries. This has also shown China's sincerity in the cooperation with the downstream countries, especially within the context where China itself was also suffering from drought condition, which has affected its household water supply and agricultural production. The Secretariat would like to extend its profound and sincere thanks to the People's Republic of China for its consideration and special arrangement of such a relief measure in the interest of downstream countries.*" When severe flooding occurred in Thailand in 2011 and Myanmar in 2015, the Chinese government was invited to send a flood prevention consulting expert group to go to the disaster area, carry out disaster reduction consultation and present a high-level consultation report, which was highly recognized by the local governments.

Since the establishment of the Joint Working Group on Lancang-Mekong Water Resources Cooperation, the work of flood prevention and disaster mitigation in the basin has been highly concerned. In the "Five-Year Action Plan for Lancang-Mekong Water Resources Cooperation" drafted by the six countries in the basin, addressing global climate change and disaster prevention and mitigation is an important part. It proposes to strengthen cooperation between the six countries in preventing floods and droughts, improve the structural facilities system, monitoring and early warning levels, and emergency support mechanisms, raise public awareness of flood and drought disasters, enhance the ability to prevent floods and droughts, reduce flood and drought losses, to ensure the sustainable economic development in this region.

5.7 Summary

This chapter investigates and analyzes flood prevention and drought relief measures in the Mekong River Basin countries, and evaluates their respective flood prevention and drought resistance capabilities. Related conclusions are as follows:

(1) Flood prevention structural measures in the Mekong River Basin countries are insufficient. Thailand has some urban revetment projects in the main urban sections of mainstream Mekong River. And there are sluices, dams (pumps) on Mekong River tributaries in Khorat. These projects can to a certain extent withstand the high water level flood from mainstream Mekong River. Laos also has some revetment works in the main urban sections of mainstream Mekong River, but most of the river banks and tributaries are in the original natural state and are susceptible to high water level floods from mainstream Mekong River. Due to its location in alpine valley, small range of river section and sparse population in the area, there is basically no flood prevention project on mainstream Mekong River in Myanmar. In Cambodia, except for a few large cities such as Phnom Penh with river revetment projects, there is basically no flood prevention project in other areas. Viet Nam has some revetment measures in the Mekong Delta. There are also some tidal gates and other facilities in the offshore area. However, these structural measures are weak and cannot withstand the high water level impact of the Mekong River.

(2) The Mekong River Basin countries have established their own disaster prevention and mitigation organizations, covering disaster prevention and mitigation departments at different levels, from the state to the province, city, and village. (The actual emergency response situation of each country at the time of the disaster is not well documented). Under the organization and coordination of the Mekong River Commission, countries in the Mekong River Basin (except for Myanmar, which is not a member country) have achieved good results in monitoring and forecasting the water conditions of Mekong River, and have a crucial role in the flood forecasting of Mekong River. However, there are no special departments for coordinating flood prevention. The flood prevention tasks are the responsibility of each region and relevant departments. For example, the irrigation department in Thailand is responsible for flood prevention in its own irrigation area. The reservoir is subordinate to power department or irrigation department. There is no consideration of flood prevention function by the reservoirs. The upstream flood regulation may cause downstream disasters, which means the flood prevention cannot be comprehensively coordinated.

(3) The overall flood prevention capacity of the Mekong River Basin countries is to be improved, and their geographical location and flood prevention status are different. From the perspective of the river basin, the countries along the Mekong River need to work and coordinate together to achieve the flood prevention effect at the basin level. For example, the mainstream Mekong River and tributary reservoirs shall be set with certain flood prevention capacity to regulate floods. The floodplain of mainstream Mekong River and Tonle Sap Lake in Cambodia have the most suitable locations and capacity for flood storage and detention. This area plays a very important role in ensuring the flood prevention security of the Mekong Delta.

(4) As far as the national level is concerned, there are certain differences in the national economic support capacity and water conservancy project support capabilities among the five countries in the Mekong River Basin, resulting in different drought relief capabilities. In the basin, the per capita GDP of Thailand and Viet Nam is more prominent, which has promoted the economic investment of the two countries in establishing emergency response systems and engineering projects for drought and disaster relief. Specifically, a series of large and medium-sized irrigation systems and projects have been built on Mekong River in Viet Nam and Thailand, forming a high-coverage irrigation network and making the irrigated area accounting for 10% or even 30% of local land, which means it to some extent possesses the ability of drought relief. The distribution of irrigation projects and irrigation districts in Cambodia is uneven, mostly concentrated in the southern areas such as the capita Phnom Penh and Tonle Sap, and there are almost no irrigation areas in other areas, resulting in significant regional differences in drought relief capacity of water conservancy structural measures. Lao PDR and Myanmar are two countries with relatively weak economies in the basin, which limit their investment in drought relief structural measures. At present, in the basin countries, the number of irrigation projects and the size of irrigation districts in Laos are behind the others. It has the lowest proportion of irrigation area (less than 1%). It should be noted that Lao PDR is expanding investment and development of hydropower projects, increasing the number of irrigation projects and the scale of irrigation districts, and has great potential in improving national drought relief capacity.

(5) At the basin level, in the process of drought relief, the upstream countries of Mekong River Basin directly draw water from Mekong River to the tributaries through diversion irrigation project for water use of local rice fields. Currently, there is a lack of coordination of emergency water between upstream and downstream countries. Although the water supply from Lancang River reservoir may relieve the hydrological drought of mainstream Mekong River to a certain extent, it has limited influence over the drought relief in Mekong Delta. Therefore, the basin countries should further strengthen coordination and communication, carry out scientific deployment and management of emergency water sources, and tap the drought relief potential of the whole basin.

(6) Working together to deal with floods and droughts has become the consensus of the countries in the basin. While building their own flood prevention and drought relief systems, all countries actively carry out international cooperation on information sharing, technical exchanges and assistance to jointly solve the problem of flood and drought in the basin. From the establishment of the Mekong River Commission to setting up of Lancang-Mekong Water Resources Cooperation mechanism, the platform for flood prevention and drought relief cooperation in the basin has become increasingly diversified and broader. The joint efforts of the countries in the basin have provided new opportunities for coordinating and coping with floods and droughts from the perspective of the whole basin.

Dr. Ding's research primarily focuses on hydro-hydraulic modeling, flood risk management, flood disaster assessment and water resources application of remote sensing and GIS. He was the Program Manager of important national and provincial level projects. The number of investigated or participated projects and programs exceeds 50. He has written 2 technology standards (or specifications) and 5 books, has published over 40 papers.

Dr. Zhang's research primarily focuses on drought monitoring and forecasting, hydrometeorology, and climate change. He has developed an operational drought monitoring and forecasting system over China and the Mekong River basin, respectively, in support of drought early-warning. To date, he has been funded as the director of Chinese Postdoctoral Science Foundation, and participated in more than 10 national-level funding projects. He has published more than 15 papers in the famous high-impact journals like *Nature Climate Change*, and also been involved in 3 published books like the *AGU Geophysical Monograph Series*.

Dr. Han's research primarily focuses on flood risk management theory, flood risk analysis. He was the program manager of important national and provincial level projects. The number of investigated or participated projects and programs exceeds 30. He has written 2 books, has published over 30 papers.

Dr. Liu's research primarily focuses on hydrological processes and modeling, flood management, and transboundary river water cooperation. She has over 5 years' research experiences in transboundary water cooperation of China. She was a participant of various Lancang-Mekong water resources cooperation activities. She is an editor of a joint publication by MWR of China and MRC in 2016. She is Program Managers of 3 international cooperation projects with Mekong River countries. She is author of over 10 journal papers and 1 book.

Chapter 6
Main Findings and Recommendations

Hui Liu and Xiangpeng Mu

Abstract Based on the investigation and analysis of the losses and basic characteristics of flood and drought disasters and development status of flood prevention and drought relief projects in the Mekong River Basin countries, and evaluation of the flood prevention and drought relief capabilities of the countries in the basin, the main conclusions are summarized in this chapter from the angle of cause, flood, drought and management. To improve the flood prevention and drought relief capability of the basin, suggestions are proposed from the following three aspects. (1) Exploit potentialities and enhance the ability of countries to cope with natural disasters; (2) Make overall plans and coordinate to enhance the disaster mitigation ability from the whole-basin perspective; (3) Strengthen cooperation and carry out technical exchanges and mechanism building.

Lancang-Mekong River is a trans-boundary river that flows through the most Southeast Asian countries and has produced splendid social culture and rich ecosystems along the coasts of China, Myanmar, Lao PDR, Thailand, Cambodia and Viet Nam. It is a natural link among six countries along the basin. Although the scale of impact of weather system is often smaller than that of the Lancang-Mekong River Basin, the geographical proximity and the natural hydraulic links of the Lancang-Mekong River Basin make the upstream and downstream and the left and right bank countries closely linked with the disaster situation, thus making the six countries in the basin face common natural disasters and water resources challenges.

Based on the flood and drought problems faced by the five countries in the Mekong River Basin, this assessment report investigated and analyzed the losses and basic characteristics of flood and drought disasters and development status of flood prevention and drought relief projects in the Mekong River Basin countries, and evaluated

H. Liu (✉) · X. Mu
China Institute of Water Resources and Hydropower Research, A-1 Fuxing Road, Haidian District, Beijing 100038, China
e-mail: wuyouliuhui@163.com

X. Mu
e-mail: swood2002@163.com

H. Liu (ed.), *Flood Prevention and Drought Relief in Mekong River Basin*, Springer Tracts in Civil Engineering,
https://doi.org/10.1007/978-981-15-2006-8_6

the flood prevention and drought relief capabilities of the countries in the basin. The main conclusions are as follows:

Cause

(1) Affected by topographic features, southwest monsoon and tropical cyclone, the Mekong River Basin has formed the spatial distribution characteristics of water resources with increasing rainfall from west to east, as well as heterogeneous flood and drought challenges.
(2) Flood in this region is usually caused by tropical cyclone and southwest monsoon, which brings heavy or lasting rainfall. Drought is usually directly related with rainfall deficit and high temperature, some extreme droughts are related with El Nino. Fresh water deficit in the delta area is a complex result rooted in strong tide, rainfall deficit and low runoff from upstream.

Flood

(3) Flood is the main form of natural disaster threatening the Mekong River Basin countries. The number of deaths caused by flood is higher in the basin, with Cambodia and Cuu Long River Delta in Viet Nam accounting for the largest proportion of deaths and northeastern Thailand and Lao PDR accounting for a relatively small proportion. Floods have a significant impact on agriculture. From the regional distribution of affected agriculture, Cambodia, Cuu Long River Delta in Viet Nam and northeastern Thailand are the main affected areas. From the perspective of economic loss in Mekong River Basin of the countries caused by floods, Cambodia has the largest amount of losses, followed by Thailand and Viet Nam, and Lao PDR with the fewest losses.
(4) There are two main types of flood in the Mekong River Basin. For the riverine flood, it could be predicted and effectively prevented from thereby damages with proper measures. For the flash flood, it is usually local and hard to predict. It is important for the local agencies to develop monitoring and early warning system, as well as carry out immediate relief and aid.
(5) Based on yearly flood peak analysis of 1985–2016, it shows that the maximum flood peak happened in 2002 and 2008 at the upper reach of Mekong River Basin (Chiang Saen, Luang Prabang and Nong Khai), with an extreme peak of around 20,000 cumec; while that of the middle reach occurred in 1991, 1996, 1997, 2000, 2001 and 2011, with an extreme peak of 70,000 cumec at Stung Treng.
(6) By defining the onset and withdrawl date of flood season, the annual flood volume and duration at mainstream stations along Mekong River was calculated and analyzed. The flood volume shows an increasing trend from upper to lower reaches, with 54 km^3 at Chiang Saen and 306 km^3 at Stung Treng. The flood duration distinguishes among years and stations, with average ranging from 128 days to 135 days.

(7) Based on flood routing analysis in the wet season of 2000, it is found that the time for Mekong River mainstream flood peak propagting from Chiang Saen station to Kratie station is about 14–15 days. As to damaging flood composition, the gauged flood data at Stung Treng station and simulation analysis results show that the contribution rate of Mun River, Sekong River and Nam Ngum to Stung Treng flood volume is the highest, 15.9%, 8.6% and 6.0% respectively; the contribution rates of Nong Khai-Nakhon Phanom, Pakse-Stung Treng and Nakhon Phanom-Pakse are higher than other regions, 25.0%, 21.3% and 19.6% respectively.

(8) The flood-prevention structural system in the Mekong River Basin countries is to be improved. Although most of the major cities along the coast have embankment or revetment works, most of the reaches can hardly withstand the impacts of high water level floods in the Mekong River. There are few large reservoirs in the basin. In Cambodia, the floodplains in the mainstream Mekong River and Tonle Sap Lake are the naturally formed locations in the basin and flood detention areas with the most suitable capacity. The existence of the flood detention area plays an important role in ensuring the flood prevention safety of the Mekong Delta.

Drought

(9) Drought has a large scope of influence, lasts for a long time and causes huge economic losses. Compared with floods, droughts occur less frequently, but cause huge losses and the total losses show a significant increase trend. From the spatial distribution of drought losses, drought losses in northeastern Thailand, Cambodia and Viet Nam are higher, while drought losses in Lao PDR are relatively low.

(10) The results of meteorological drought analysis show that the drought severity in northeastern Thailand, most of Cambodia and Myanmar has increased over the past half century, especially in some parts of northeastern Thailand; in most parts of the basin, the frequency of meteorological drought is close to 25% due to low rainfall, especially in northeastern Thailand and Cambodia; and Cambodia and Viet Nam Delta in the lower reaches are more liable to severe and exceptional droughts. The results of hydrological drought analysis show that there is no obvious drought at Chiang Saen station and Mukdahan station after 2005; while the downstream Stung Treng station experienced more severe droughts in 2010 and 2016 (SRI approaching −2). By analyzing the causes of typical drought events, it is found that the main reason for the drought in the Mekong River Basin is the extraordinarily less rainfall (compared with that in the same period in history) caused by El Nino and anomalies of atmosphere-ocean circulation system.

(11) Because of the differences between the national economic supporting capacity and development of water conservancy projects, the drought resistance ability of different countries is different. Among them, Viet Nam and Thailand

have built high-density, large-scale irrigation projects and irrigation network, providing a reliable engineering guarantee for the development of drought resistance. The distribution of irrigated areas is uneven in Cambodia and the ability of drought resistance in different regions is different. The number of irrigation projects and irrigation area in Lao PDR is relatively small and drought resistance ability is slightly inadequate.

Management

(12) The Mekong River Basin countries have established their own organization system for disaster prevention and reduction, covering different levels of disaster prevention and reduction departments from the state to provinces, cities and villages. Under the organization and coordination of the Mekong River Commission, the countries in the basin have carried out better monitoring, forecasting and early warning of the water regime of the mainstream Mekong River. However, all countries, except Thailand, do not have specific flood prevention departments, so it is difficult to carry out unified flood prevention command, regulation and decision-making. At the basin level, with regard to emergency water for drought resistance, the countries are still in the state of giving priority to their own needs at present. Although there is joint drought relief cooperation among countries in the basin, the scientific allocation and management of water resources have not been carried out from the perspective of the whole basin.

(13) Working together to deal with flood and drought disasters has become the consensus of all countries in the basin and joint efforts have provided a new opportunity for overall and coordinated response to flood and drought disasters from the perspective of the whole basin. While building their own flood prevention and drought relief systems, all countries actively carry out international cooperation in information sharing, technical exchange and assistance to jointly solve the flood and drought problems in the basin. From the establishment of the Mekong River Commission to the establishment of the Lancang-Mekong Water Resources Cooperation, the platform for basinwide flood prevention and drought relief cooperation has become increasingly rich and broad.

Recommendations

(1) Exploit potentialities and enhance the ability of countries to cope with natural disasters.

The Mekong River Basin countries have established organization system for disaster prevention and reduction to deal with flood and drought disasters. Although there is a significant correlation between the national flood prevention and drought relief capacity and economic and social development level, the consensus of the whole basin on improving the flood prevention and drought relief structural measures is also very important. Based on the viewpoint of engineering, Thailand and Viet Nam have taken adequate measures to combat drought. Lao PDR has more reservoirs established, under construction and planning. Cambodian drought-resistant potential needs to be tapped and developed.

(2) Make overall plans and coordinate to enhance the disaster mitigation ability from the whole-basin perspective.

The reservoirs built in the Mekong River Basin have a capacity of more than 20 billion m^3. It is suggested that the possible positive role of these reservoirs in flood prevention and drought relief in all countries and the whole basin should be studied in depth, and the impacts of flood prevention and drought relief projects on the water disasters in the downstream regions or countries should be deeply analyzed while they alleviate the local disasters to lay a foundation for the overall coordination of the basin.

(3) Strengthen cooperation and carry out technical exchanges and mechanism building.

The establishment of Lancang-Mekong Cooperation Mechanism provides a new platform for China and five other countries to exchange deeply and share their experience in water resources management and jointly address the challenges of climate change. The theme spirit of "Shared River, Shared Future" is China's appeal for cooperation in the whole basin, and also expresses a profound understanding of the Lancang-Mekong water resources cooperation. In addition to the existing bilateral cooperation and dialogue and cooperation mechanisms, it is suggested that the six countries in the basin should strengthen the technical exchanges and cooperation in flood prevention and drought relief under the Lancang-Mekong Cooperation Mechanism and study the possibility of establishment of a coordinated mechanism for flood prevention and drought relief in the whole basin to work together for the sustainable development of the basin and equitable and rational utilization of water resources.

(4) Proposed cooperation projects and activities

Capacity building on water modelling tools and disaster mitigation related technologies are recommended to be carried out by all 6 countries under LMC mechanism.

Limitations of this assessment report

Due to the limitation of data and research time, the statistical data of disaster losses in this project may not be comprehensive and accurate; the accuracy of flood composition analysis is likely to be affected because it is based on statistical analysis rather than observed hydrological data, and the contribution rate of tributaries and regions to floods is based on overall statistical analysis and is not analyzed in detail according to different types of floods; some of the flood prevention and drought relief projects are based on manual interpretation of remote-sensing images and may be missed or inaccurate. In conclusion, this research analyzed the main flood and drought characteristics of the Mekong River Basin and qualitatively assessed the overall disaster prevention and mitigation capacity of the basin.

Dr. Liu's research primarily focuses on hydrological processes and modeling, flood management, and transboundary river water cooperation. She has over 5 years' research experiences in transboundary water cooperation of China. She was a participant of various Lancang-Mekong water resources cooperation activities. She is an editor of a joint publication by MWR of China and MRC in 2016. She is Program Managers of 3 international cooperation projects with Mekong River countries. She is author of over 10 journal papers and 1 book.

Dr. Mu's research primarily focuses on water resources management, river ecosystem restoration and hydraulic computation. He has over 10 years' research experiences in relative national level projects. He is author of over 40 journal papers and 4 books, as well as dozens of patents for inventions.

Appendix
Introduction of THREW Model

Fuqiang Tian and Shiyu Hou

I. **Principle**

Only the principle of mass conservation is taken into account in conceptual hydrologic models, physical hydrologic models are evidently more advantageous because the principles of energy conservation and mass conservation are both taken into account, underlying surface conditions and meteorological factors are utilized fully and more detailed distributed modeling is conducted, and the physical significance of hydrologic model parameters are improved. Nevertheless, because underlying surface is highly changeable, spatial discretization leads to an excess of calculation elements and over-parameterization, and different parameter schemes can reach the same simulation result, namely the phenomenon of "same effect of different parameters". Physical hydrologic models are also faced with the problems that underlying surface conditions and meteorological conditions are hard to access, calculate and verify in data-insufficient areas. The fundamental reason lies in the mismatching between the applicable scale of the mathematical physical equation of hydrologic modeling and the model's applied scale. For instance, point scale or representative unit volume scale is applicable to Richards equation, Boussinesq equation and Saint-Venant equation, but the parameters, initial conditions and boundary conditions needed cannot be observed using current hydrologic methods, so parameters can be figured out only through calibration, which is bound to cause over-parameterization.

Reggiani proposed a hydrologic simulation method based on representative elementary watershed (REW) in 1998. In this method, a basin is set apart into REW and then, divided into several functional sub-regions. Subsequently, mass, momentum and energy conservation and equilibrium equations on the micro scale are established in the functional sub-region and after homogenization, a system of ordinary

F. Tian (✉) · S. Hou
Tsinghua University, New Hydraulic Building, Beijing 100050, China
e-mail: tianfq@mail.tsinghua.edu.cn

S. Hou
e-mail: housy16@mails.tsinghua.edu.cn

H. Liu (ed.), *Flood Prevention and Drought Relief in Mekong River Basin*, Springer Tracts in Civil Engineering,
https://doi.org/10.1007/978-981-15-2006-8

differential equations on the REW's scale is established. Because the equation system is unclosed, ordinary differential equations of geometrical relationship and constitutive relationship are added to make the equation system closed. However, the REW concept Reggiani et al. proposed was not clearly and scientifically defined. In the model, there is no ice or snow module, vegetation is not taken into account, simulation effect of high mountains, cold regions and other regions with a big ratio of snow and ice is unsatisfactory, and evaporation and transpiration cannot be distinguished from each other.

Therefore, Tian Fuqiang et al. redefined REW in a rigorous manner by dividing REW into surface and subsurface layers and dividing subsurface layer into 2 sub-regions and surface layer into 6 sub-regions. The subsurface layer is divided into saturated and unsaturated zones by groundwater level. The saturated zones are below the groundwater level and unsaturated zones are above the groundwater level and below the surface. In order to consider the freezing and thawing of water in soil, saturated and unsaturated zones, besides soil, liquid water and gas, should contain ice phase substances.

In the surface layer, the most notable feature is that the stream network and its adjacent hillsides form an interconnected system. From the point of view of hydrological simulation, the hillsides and stream network are the two basic components of the basin.

(1) The stream network is further divided into main channel and sub-stream network. In order to maintain the geometrical invariability of REW, the lakes, reservoirs, streams and channels on the sub-REW scale are included in the sub-stream network.
(2) The hillside is the main area of precipitation redistribution and transportation, which is divided into four zones: bare soil zone, vegetated zone, snow covered zone and glacier covered zone.

Finally, the REW is divided into surface and subsurface layers. The subsurface layer is further divided into 2 sub-regions and surface layer is further divided into 6 sub-regions, which can reflect four typical underlying surface types: bare soil, vegetation, snow, glacier, as shown in Fig. A.1.

In the horizontal direction, the 6 sub-regions of the surface layer form a complete coverage of the surface. In the vertical direction, bare soil zone, vegetated zone, snow covered zone and glacier covered zone are located at the unsaturated zone; according to the relationship with the groundwater level, other two surface sub-regions (sub-stream network and main channel) may be located at the saturated zone, or at the unsaturated zone. According to the availability of data and purpose of application, a specific spatial scale is selected to set apart the basin into REW, thus making the attributes of each REW including soil, vegetation, snow and glacier uniquely determined.

In the surface sub-regions, surface runoff is generated when the precipitation intensity exceeds infiltration capacity. In unsaturated zone, soil heterogeneity can result in subsurface flow. For example, if the unsaturated zone is divided into upper and lower layers and the upper layer's infiltration capacity is greater than that of the

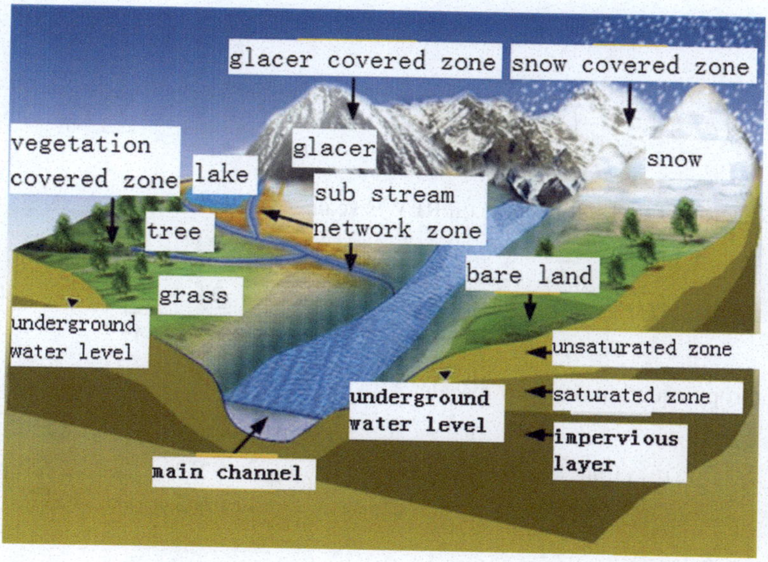

Fig. A.1 REW zoning in THREW model

lower layer, the subsurface flow may take place at the interface of two layers of soil. In order to simplify the calculation, the heterogeneity is not directly considered in the definition of REW and the subsurface flow and preferential flow can be expressed by material exchange items and corresponding constitutive relationship between unsaturated zone and adjacent REW as well as outside basin.

In order to consider evapotraspiration, the surface sub-regions contain vapor phase substances. The soil pores in unsaturated zone contain air and water vapor, which are considered as gases and defined as a phase substance. Similar phenomena are found in the snow covered zone. The water storage in the bare soil zone can be regarded as depression detention water and water storage in the vegetated zone can represent the sum of the water quantity of vegetation interception and depression detention. The runoff from surface sub-regions flows into the sub-stream network, and then into the main channel.

II. Basic Equations

In the THREW model, basin is divided into sub-systems at REW level, sub-region level and phase material level. It leads to the general form of conservation laws on REW's scale.

(1) Equation of mass conservation on REW's scale

$$\frac{d}{dt}\left(\overline{\rho_\alpha^j}\varepsilon_\alpha^j y^j \omega^j\right) = \sum_{P=EXT,L,T,B,i}^{L=1\ldots N_k, i\neq j} e_\alpha^{jP} + \sum_{\beta\neq\alpha} e_{\alpha\beta}^j \tag{A.1}$$

(2) Equation of momentum conservation on REW's scale

$$\left(\overline{\rho_\alpha^j}\varepsilon_\alpha^j y^j \omega^j\right)\frac{d}{dt}\left(\overline{v_\alpha^j}\right) = \overline{g_\alpha^j \rho_\alpha^j \varepsilon_\alpha^j y^j \omega^j} + \sum_{P=EXT,L,T,B,i}^{L=1\ldots N_k, i\neq j} T_\alpha^{jP} + \sum_{\beta\neq\alpha} T_{\alpha\beta}^j \tag{A.2}$$

(3) Equation of thermal balance on REW's scale

$$(\varepsilon_\alpha^j y^j \omega^j c_\alpha^j)\frac{d\overline{\theta_\alpha^j}}{dt} = \overline{h_\alpha^j \rho_\alpha^j \varepsilon_\alpha^j y^j \omega^j} + \sum_{P=EXT,L,T,B,i}^{L=1\ldots N_k, i\neq j} Q_\alpha^{jP} + \sum_{\beta\neq\alpha} Q_{\alpha\beta}^j \tag{A.3}$$

(4) Entropy equation on REW's scale

$$\left(\overline{\rho_\alpha^j}\varepsilon_\alpha^j y^j \omega^j\right)\frac{d\overline{\eta_\alpha^j}}{dt} = \overline{b_\alpha^j \rho_\alpha^j \varepsilon_\alpha^j y^j \omega^j} + \overline{L_\alpha^j}\varepsilon_\alpha^j y^j \omega^j + \sum_{P=EXT,B,T,L,i}^{L=1\ldots N_K, i\neq j} F_\alpha^{jP} + \sum_{\beta\neq\alpha} F_{\alpha\beta}^j \tag{A.4}$$

III. **Constitutive Relationship**

The system of said conservation equations is indeterminate for the number of variables is bigger than that of equations, and other relational expressions should be added for solving the equations. Therefore, geometrical and constitutive relationships must be added to make the equation system closed.

In hydrology, the linear relationship, namely a constitutive relationship, of Darcy's law is true in a given scope of flow velocity. However, constitutive relationship is less accurate than kinematic formula. In THREW model, formulas are established on REW's scale, and traditional constitutive relationship on micro scale, such as Chezy formula, Darcy's law and soil moisture characteristic curve are no longer applicable. Up-scaling of constitutive relationship is mainly realized with statistical physics method.

In the model, equations are further simplified, and momentum conservation equations of different sub-regions are closed. Then, mass exchange term, vegetation interception and other constitutive relationships and the 1 area ratio of saturated zone and other geometrical relationship are established to finish the establishment of the closed system of equations.

IV. **Model Application**

THREW has been applied in the basins with different climatic and hydrological conditions, including the Lancang River, including humid region, semi-arid region, alpine valley region and so on. Table A.1 lists some of the THREW model-applied basins.

Table A.1 List of THREW model-applied basins

Country	Basins	Basin characteristics	Published papers or THREW model-applied projects
China	Lancang River	160,000 km^2, alpine and gorge region	Applied in fund projects research
China	Tailan River	13,240,000 km^2, Tianshan region, glacier	He et al. [1], HESS
China	Three Gorges	56,000 km^2, monsoon, complex terrain	Applied in the entrusted projects of Three Gorges Corporation
China	Ya-lung River	128,000 km^2, alpine and gorge region	Applied in fund projects research
China	Hanjiang River Basin	95,000 km^2, humid mountain basin	Sun et al. [2], J. Hydrol.
China	Urumqi River	289,000 km^2, permafrost region	Mou et al. [3], HESS
China	Chabagou River	1,860,000 km^2, arid Loess Plateau	Applied in fund projects research
China	Weihe River	25,000 km^2, arid Loess Plateau	Liu et al. [4], HSJ
China	Jinghe River	4500 km^2, arid Loess Plateau	Applied in fund major projects research
China	Chaohe River	4808 km^2, semi-humid earth-rock mountain region	Applied in fund major projects research
China	Baihe River	8603 km^2, semi-humid earth-rock mountain region	Applied in fund major projects research
China–India	Yarlung Zangbo-Brahmaputra River	520,000 km^2, upper ice and snow, the lower monsoon	Applied in the entrusted projects of Lancang River Corporation
China-Kazakhstan	Yili River	90,000 km^2, trans-boundary water, drought	Applied in the entrusted researches of China Renewable Energy Engineering Institute

(continued)

Table A.1 (continued)

Country	Basins	Basin characteristics	Published papers or THREW model-applied projects
China-Kazakhstan	Emin River	20,000 km^2	Applied in the entrusted researches of China Renewable Energy Engineering Institute
China-Kazakhstan	Eerqisi River	89,000km^2, trans-boundary water, drought	Applied in the entrusted researches of China Renewable Energy Engineering Institute
The United States	Sangamon	3150 km^2, agricultural watershed	Liu et al. [5], IAHS Publ.
The United States	Reynolds	238.9 km^2, mountain area, snow	Applied in fund projects research
The United States	Salt Creek	49.9 km^2, urban watershed	Applied in the project research of American Foundation
The United States	Blue River	1233 km^2, temperate continental climate	Li et al. [6], J. Hydrol.
The United States	Illinois river	2484 km^2, temperate continental climate	Li et al. [6], J. Hydrol.
Lao PDR	Nam Ou River	3684 km^2, tropical monsoon climate	Applied in the entrusted researches of Kunming Engineering Corporation Limited
Austria	Lienz	1198 km^2, Alps	He et al. [7], HESS

References

1. He ZH, Tian FQ, Gupta HV et al (2015) Diagnostic calibration of a hydrological model in a mountain area by hydrograph partitioning [J]. Hydrol Earth Syst Sci 19(4):1807–1826
2. Sun Y, Tian F, Yang L et al (2014) Exploring the spatial variability of contributions from climate variation and change in catchment properties to streamflow decrease in a mesoscale basin by three different methods [J]. J Hydr 508:170–180
3. Mou LQ, Tian FQ, Hu HP (2009) Artificial neural network model of runoff prediction in high and cold mountainous regions: a case study in the source drainage area of Urumqi river [J]. J Hydr Eng 1:64–69 (in Chinese)
4. Liu DF, Tian FQ, Hu HC et al (2012) The role of run-on for overland flow and the characteristics of runoff generation in the Loess Plateau, China [J]. Hyd Sci J 57(6):1–11
5. Liu DF, Tian FQ, Hu HP (2009) Sediment simulation at Upper Sangamon river basin using the THREW model [J]. IAHS-AISH Publication 335:187–195

6. Li HY, Sivapalan M, Tian FQ (2012) Comparative diagnostic analysis of runoff generation processes in Oklahoma DMIP2 basins: the Blue river and the Illinois river [J]. J Hyd 418:90–109
7. He ZH, Parajka J, Tian FQ, Bloschl G (2014) Estimating degree-day factors from MODIS for snowmelt runoff modeling [J]. Hydrol Earth Syst Sci 18(12):4773–4789

Dr. Tian's research primarily focuses on hydrological processes and modeling, flood management, agricultural water management, and transboundary river water management. He is editor of two journals of *Hydrology and Earth System Sciences* and *Journal of Hydrology*. He is the Chair of Panta Rhei Initiative (the IAHS Scientific Decade 2013–2022) for the fourth biennium (2019–2020), and also the chair of Biofuel Working Group of ICID from 2015 to the present. He has published over 150 journal papers, 3 books and 1 textbook.

Ms. Hou's research primarily focuses on hydrological processes, hydrological predictions and modelling, and flood management. She is in Department of Hydraulic Engineering of Tsinghua University. She has published over 1 journal papers.

The manufacturer's authorised representative in the EU is Springer Nature Customer Service Centre GmbH, Europaplatz 3, 69115 Heidelberg, Germany. If you have any concerns regarding our products, please contact ProductSafety@springernature.com

Printed and bound by CPI Group (UK) Ltd, Croydon, CR0 4YY
23/09/2026
02221542-0001